Contracting for sustainability

Contracting for sustainability

An analysis of the
Lake Victory-EU Nile perch chain

Emma Verah Kambewa

Wageningen Academic
P u b l i s h e r s

This book was originally written as PhD thesis with the title:
Balancing the people, profit and planet dimensions in international marketing channels
A study on coordinating mechanisms in the Nile perch channel from Lake Victoria

ISBN: 978-90-8686-037-1

First published, 2007

Wageningen Academic Publishers
The Netherlands, 2007

Abstract

Sustainable development hinges on a combined focus of its impact on society (people), the environment (planet) and to its economic value (profit). Increasingly, it is being recognized that these people, profit and planet dimensions are interlinked and an important challenge for public and private policy is to take them jointly into account. This inter-linkage is particularly evident in international channels that build on scarce natural resources from developing countries, which is the focus of this thesis. Specifically, we focus on how international marketing channels can be organized in order to enhance the balance between the people, profit and planet dimensions of the sustainable development such that small-scale primary producers from developing economies are integrated into the international marketing channels in a way that adds to the profitability of the channel and the welfare of the local communities, without compromising the sustainability of natural resources. We focus on contracts as mechanism to stimulate welfare, sustainability and quality at primary stages of the channel. We apply a number of theoretical approaches namely transaction costs economics, social – and network theory, and property rights theory to develop and test arguments for contracts to stimulate sustainable and quality-enhancing practices. We undertook both case study and conjoint analysis to develop and test our arguments.

A situational analysis of the context in which primary producers operate lead to a conclusion that primary producers (i.e., fishermen) fail to implement sustainable and quality-enhancing practices because of major bottlenecks that they face such as the degradation of natural resources (i.e., fisheries), limited access to production facilities, information asymmetries and ineffective enforcement. An empirical analysis at fishermen and middlemen levels shows that they are open to contracts as mechanisms to stimulate sustainable and quality-enhancing practices. This is true especially if such contracts provide production facilities, price information, bring fishermen closer to international channels and allow private policy enforcement of sustainable practices. However, both fishermen and middlemen have idiosyncratic preferences for the particular types of contracts. This implies that they should be offered a choice among different contracts. An analysis of channel members in the downstream part of the channel show that in order to create a situation in which fishermen and middlemen are engaged in sustainability and quality-enhancing contracts, downstream channel members and other stakeholders should be engaged in and/or support micro-projects that may enable fishermen to solve the market failures that they face. The channel members farther downstream however may need stakeholder pressure, e.g. from special interest groups to be involved in addressing the market failures in the upstream.

Key words: people, profit, planet, marketing channels, contracts, sustainability, quality, Lake Victoria, conjoint analysis

Table of contents

Acronyms

BMU	Beach Management Unit
CPR	Common property resources
CSR	Corporate Social Responsibility
EU	European Union
FAO	Food and Agriculture Organization
FD	Fisheries Department
GTZ	Deutsche Gesellschaft für Technische Zusammenarbeit
HACCP	Hazard Analysis of Critical Control Points
IUCN	The World Conservation Union
KMFRI	Kenya Marine and Fisheries Research Institute
LVFO	Lake Victoria Fisheries Organization
MSC	Marine Stewardship Council
MSY	Maximum Sustainable Yield
NR	Natural Resources
PPP	People, Profit and Planet
TCE	Transaction cost economics
UNICEF	United Nations Children's Fund
WOTRO	The Netherlands Foundation for the Advancement of Tropical Research
WWF	World Wildlife Fund

Acknowledgements

Different people - numerous to mention one by one contributed in different ways to the success of this study. I hereby mention just a few. First and foremost, the support, encouragement and love that I cherished from my family during the entire four years cannot be equated to any quantitative measure. My husband, Daimon – your faithfulness and trustworthiness inspired me to take leave of absence from home. Your tolerance of my four - year absence and perseverance and love to take care of our children shall never depart from my heart, mind and soul. My daughter, Tonthozo and son, Lonjezo, although you were respectively 8 and 4 years at the time I left, you accepted and endured my four-year absence with maturity and love – a unique character at your age. For you, I will ever be a caring and loving mother you have ever known and will support you as you map out your destiny.

I would not have had a scholarship if it were not for Prof., Dr. Ruerd Ruben. You helped me to get this chance which truly deserves my profound appreciation. I also thank Dr. Daniel Jamu for first encouraging and inspiring me to aim higher and also for supporting me throughout my study.

This study started and finished in time, thanks to the effort and support from my excellent supervision team. I sincerely wish to thank my promoters, Prof. Dr. ir. J.C.M van Trijp and Prof. Dr. ir. M.A.J.S. van Boekel, for the untiring and foresighted guidance throughout the study. I also wish to thank my co-promoters Dr. Aad van Tilburg and Dr. Paul Ingenbleek for the immeasurable day to day support during my study. As my promoters and co-promoters, you always left your office doors open that I could walk in and interrupt you anytime. This was a unique kindness never to be taken for granted especially in the land where appointments are the normal trend to meet (busy) people. I also whole heartedly wish to thank Dr. Ivo van der Lans for never allowing me to despair with the technical conjoint analysis issues beyond what I could persevere. You always had an answer to all my questions no matter how vague they (sometimes) were. I never took your kindness for granted. Special thanks also to Dr. Ruud Verkerk for his contribution to my work especially on issues relating to quality.

A special appreciation to Aad and your wife, Augustine, you took me as your child. You kept your door open for me to come in and feel at home far away from home anytime I needed parental love and encouragement. You could never be better parents to me in a foreign land than you indeed were.

Working within the corridors of the Marketing and Consumer Behaviour Group (MCB) was never a mistake thanks to you colleagues and friends – Amber, Janneke, Timon, Vladimir, Frans, Erno, Ellen van Kleef, Erica, Eric, Kalaitzis, Margreet; Marcel, Meike, Jantine, Filip, Ynte, Arnout, Lynn, Olivia, Judith, Heleen and Tineke; those that left the MCB – Wendy, John, Clara and my officemates, Cynthia, Ornella and Jean-Francois. Your support and encouragement during my intermittent four years of being among you in MCB was great. To

Ellen Vossen and Liesbeth, you made my stay in Holland and work in MCB a lot easier because you always showed me the way to find anything I needed.

The contributions and encouragement from colleagues in the WOTRO program cannot go unacknowledged. Discussions and comments during the meetings helped me to be focused. Lu and Zuniga - I will always cherish the fond memories we shared together especially sharing our academic experiences as well as traditional meals. Wishing you success.

This study would never have come true if it were not for the financial support from The Netherlands Foundation for the Advancement of Tropical Research (WOTRO) and hospitality of the Kenya Marine and Fisheries Research Institute (KMFRI). I thank the management of the two institutions most profoundly. A special word of appreciation goes to the Deputy Director – Dr. Enock Wakwabi and the Assistant- Dr. Richard Abila – of KMFRI for accepting to host me and the support rendered to me in every respect during my 1.5 years of stay in Kenya. The entire staff at KMFRI was wonderful. I could never have collected meaningful data if it were not for the hard working spirit of Robert, Veronica and Patrick for persevering the scorching heat along the equator and all the drivers that safely negotiated the rough roads to the field. More importantly, the jokes and stories we shared made me know Kenyan life and culture better. You are a wonderful team. Special thanks to all fishermen, middlemen, processors, importers, retailers and other organizations that provided the information that has made this study worthwhile.

I appreciate the nice time I had with Milcent, Evans, Lena and John - that made me live like a Kenyan. Special thanks to Mrs. Abila for making me never wanting to leave my "House" although I had to. You were the best land lady I could ever imagined in a foreign land. I also appreciate the assistance and friendship I had with Claris and her family who made my trips to Nairobi enjoyable. The support from all colleagues at KMFRI – Nairobi office was great.

I wish to thank all too many friends and colleagues from Africa and in particular, Mose, Patricia, Bongani, Enock, Chris, Bernard, Austin, MacDonald just to mention a few for the wonderful time we shared since we met in the foreign land. The Malawian team scattered in the different institutions in Holland – networking and talking in our Malawian tongue through emails, phones and in person truly reflected Malawi as a warm Heart of Africa.

Jacobus Verheul and your parents, brothers and sisters, you made me know how to celebrate Christmas in Holland. I will never forget that experience. Laurentius and Jacobus, I will always remember the time we had together in Kenya eating "Ugali".

Surely, I have missed names of colleagues and friends that I should have certainly mentioned. Nonetheless I appreciate the contribution all of you made to my stay and work in Holland, Kenya as well as Malawi. May God bless you all.

Chapter I

General introduction

1.1 Introduction and research objective

Poverty and inequality continue to be widespread in developing economies (Wade 2004; Rigg, 2006; 1998), even so, at a time when globalization and integration of world markets offer opportunities for economic development (Bardhan, 2006; Thorbecke and Nissanke, 2006). Poor small-scale producers in developing economies are increasingly being excluded and marginalized from global marketing channels (Nissanke and Thorbecke, 2006; Van de Meer, 2006). Yet, increased globalization and market integration affect the local ecosystems on which they depend for their livelihoods (Aggarwal, 2006).

Marketing and related literature acknowledge that increasing poverty and degradation of natural ecosystems are some of the major challenges and criticisms of the existing marketing systems (Wilkie and Moore, 1999; Basu, 2006). For example, whereas Wilkie and Moore (1999), in their article "Marketing's Contribution to Society" acknowledge that marketing has made unprecedented contributions to economic development, they question whether the marketing system represents "the best of all worlds," i.e., whether the benefits cut across the whole society. They also question the extent to which the marketing system would be responsible to protect public interests or act as a steward of the society's resources. Similarly, Nissanke and Thorbecke (2006) question whether the present form of market integration is conducive to a growth and structural transformation process that is capable of engendering and sustaining pro-poor economic growth and favourable distributional consequences. Nonetheless, Wilkie and Moore (1999) remain optimistic that marketing can still "literally change the world" especially for the citizens of developing economies. Such change would require a balance between people, profit and planet (PPP) dimensions of the Sustainable Development (Brundtland, 1987); such that small-scale primary producers from developing economies are integrated into the international marketing channels in a way that adds to the profitability of the channels, the welfare of the local communities, without compromising the sustainability of natural resources (NR). This thesis investigates how international marketing channels can be organized in order to achieve such better balance of the PPP dimensions.

This chapter is organized as follows: In section 1.2, a brief background of this research and the case study channel is given. Section 1.3 puts sustainability of NR, food quality and people's welfare into the perspective of sustainable development. The section highlights the importance of integrating small-scale primary producers into global marketing channels in a manner that enhances the PPP dimensions at primary level and the channel as a whole. Section 1.4 introduces sustainability and quality from the context of a hierarchy of human needs. Sustainability is further elaborated in section 1.5 which argues that existing marketing mechanisms may not guarantee sustainability and hence, alternative mechanisms may be

Chapter 1

needed. Section 1.6 outlines the approach for this study and section 1.7 gives the structure of the thesis which concludes the chapter.

1.2 Research background

1.2.1 A study within a multidisciplinary program

This study comes within the framework of a multidisciplinary program on globalization, food quality and sustainable international agro-business chains. The program is coordinated by the Wageningen University and Research Center, Department of Social Sciences and funded by the Netherlands Foundation for the Advancement of Tropical Research (WOTRO). The program focuses on perishable products namely; fish, vegetables and fruits from Kenya, China and Costa Rica, respectively. The studies address the design and management of international agro-food chains in order to improve incentives for quality control and sustainable utilization of NR. This research focuses on fresh fish channel from Lake Victoria, Kenya, in particular to the EU.

1.2.2 The Lake Victoria Fresh Nile perch channel to Europe

Lake Victoria, the second largest fresh water body in the world, is in East Africa shared among Kenya, Uganda and Tanzania. Nile perch (*Lates niloticus*) - the case study fish, was introduced into the Lake in the mid 1950s to improve the productivity and commercial value of the fishery (Geheb, 1997). About two decades later, Nile perch production boomed, triggering unprecedented socio-economic benefits. For example, employment in fisheries and support sectors more than doubled between 1980s and 1990s (Abila and Jansen, 1997). About 80% of the fishermen earned primary income from fishing and the income was fairly and evenly distributed (Henson and Mitullah, 2003). Food security and nutritional status in the region improved (Geheb eds., 2002). By mid 1990s, the bordering states became net fish exporters (Bokea and Ikiara, 2000).

However, as socio-economic benefits continued to flourish, ecological problems emerged. By mid 1990s fish catch and species diversity rapidly declined. The decline has been attributed to uncontrolled fishing that erupted following increased export demand for Nile perch. For example, fishermen and fishing vessels more than tripled by 2000 (Jul-Larsen *et al.*, 2003). The decline in biodiversity and production is also attributed to habitat degradation, i.e., siltation, aquatic weeds, pollution and predation by Nile perch (LVFO, 1999).

Following the production decline, competition for fish between export and domestic markets intensified. Consequently, domestic fish consumption declined leading to increased food insecurity, malnutrition and high levels of absolute poverty along the lakeshore areas – the hub of fishing and fish processing (Geheb, eds. 2002; Bokea and Ikiara, 2000). To-date, the domestic market relies on juveniles and by-products such as skins, fats and skeletons from the factories. Some factories close down due to inadequate fish supply. At regional level less than 60% of installed processing capacity is being utilized (Henson and Mitullah, 2003).

From the international market, the EU - the main importer taking up about 80% of Nile perch export volume (see, www.globefish.org), banned Nile perch imports for three consecutive times between 1997 and 1999 due to food safety concerns (Henson and Mitullah, 2003). Although the channel strived to comply with EU food safety regulations, lack of modern technological tools for quality improvement remains a major problem in the primary stages.

Whereas regional efforts to minimise environmental pressure, over-fishing and destructive fishing are underway, success faces major challenges. The limited alternative sources of livelihood imply that rural communities exert more pressure on the fishery for survival such as relying on juvenile Nile perch. Without modern quality improvement facilities, primary producers cannot ensure quality of the fish and hence cannot benefit from global markets. In a nutshell, the channel is one of the examples that amplify the need and urgency to protect NR in a manner that enhances the welfare of the local communities.

1.3 Welfare, sustainability and quality in the context of sustainable development

This section puts people's welfare, sustainability and food quality into the perspective of sustainable development. According to the World Commission on Environment and Development (Brundtland, 1987), development is sustainable if it meets the needs of the present generation without compromising the ability of future generations to meet their own needs. That way, sustainable development requires a combined focus on society (people), economy (profit) and environment (planet) - denoted as the three pillars of sustainable development. Building on this framework, this study investigates how small-scale primary producers in developing economies can be integrated into global marketing channels in a manner that enhances the PPP dimensions in the channel (Figure 1.1).

People: The focus of this study is on the small-scale primary producers and rural communities in developing economies. The welfare of these people – who constitute the bulk of the population - is critical for the economic development of developing nations. A healthy and productive population is needed for economic development of any society. Often poor rural communities in developing nations rely on NR such as fish for their food and money to meet

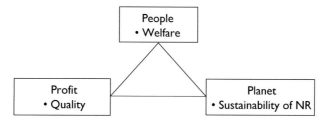

Figure 1.1. Sustainable Development model, (Adapted from Brundtland, 1987).

other necessities. Consequently, protection of NR would directly contribute to the welfare of such poor people through sustainable food (e.g., fish) production.

Profit: Of particular interest in this study is the profitability of the activities of the small-scale primary producers in developing economies and the channels they belong to. Small-scale producers constitute the bulk of primary producers in developing nations. As such, they replenish most of the international marketing channels that originate from developing economies. Hence, better integration into these marketing channels would contribute to the profitability of their activities which in turn would enhance their welfare and that of their families and communities. It would also contribute to the profitability of the overall channel they belong to. However, to be competitive, primary producers need to meet the high quality standards set by the global markets. This study focuses on how to improve the quality of products from small-scale primary producers in order for them to competitively participate in global marketing channels.

Planet: The importance of NR (such as fisheries) and the need to promote their sustainability for the economic development of developing countries cannot be overstated (see Delgado *et al.*, 2003; www.fao.org). At the same time, poverty and food insecurity in the developing economies undeniably contribute to the degradation of the NR. This study focuses on how to promote sustainability of NR, i.e., fisheries. Considering the enormous welfare and socio-economic constraints that primary producers face in developing economies, this study investigates how they can be supported and motivated to implement sustainable practices.

Increasingly, corporate social responsibility (CSR) (Brown and Dacin, 1997) is seen as a better strategy to solve some of the welfare and socio-economic problems that poor rural communities face in developing economies (Vachani and Smith, 2004; 2006; Prahad and Hart, 2002; Bhattacharya, Smith and Vogel, 2004) including environmental degradation (Bansal and Roth, 2000). In order to contribute to this, some major food processing companies (e.g., Unilever, Nestle) through the Sustainable Agriculture Initiative (SAI) (www.saiplatform.org) encourage and support competitive agricultural systems that not only protect natural environments but also improve the welfare of local communities. This study also investigates how downstream channel members and other stakeholders can help to solve the welfare, socio-economic and sustainability problems in their upstream channels.

1.4 Sustainability and food quality in the perspective of hierarchy of human needs

This section puts sustainability and food quality into the context of human needs. The section adapts the Maslows hierarchy of needs (Maslow, 1943) to place welfare, sustainability and quality among the hierarchy of human needs at different levels of abstraction (Figure 1.2).

Figure 1.2. Hierarchy of human needs in relation to welfare, sustainability and food quality (adapted from van der Hoff and Roozen, 2001).

The basic need for any human being is that of food security –access to adequate food all the time. Although food security is a basic necessity for human life and development, not all people have access to adequate food all the time. In developing economies, the majority of poor rural population tends to suffer from chronic or periodic food shortages. Unless such people meet their basic food needs, they may never attain the needs at higher levels.

Following food security is food safety. Food science and quality management literature defines food safety from the perspective of having no immediate danger of infection and poisoning from the presence of pathogenic organisms or chemical contamination (Luning, Marcelis and Jongen, 2002). Over the past decade, the incidences of food borne diseases (e.g. mad cow) and chemical contamination have heightened concerns about food safety in food supply chains. Although food safety may be at a slightly higher level of abstraction for the absolute poor people, it is a bottom line factor for those that can choose which food to eat.

Beyond food safety is food quality. Food quality takes a more personal and subjective perspective because it may be context specific such as; situation under which food is consumed; the practical, functional, ethical and emotional benefits consumers want to derive from the food; and the evolving characteristics of the consumer such as changing life styles, eating habits and demographic conditions (Leeflang and van Raaij, 1995; see also the International Journal of Research in Marketing 12, 1995). Being more subjective, food quality encompasses different attributes such as taste, flavour, price, freshness, texture, nutritional value and convenience among others depending on the product. An evaluation of these personal perspectives of quality defines a consumer perception of quality that reflect in general terms the superiority of the product in meeting intended use (Zeithaml, 1988; Luten *et al.*, eds., 2003). Consumers' endless pursuit of these quality attributes creates a market climate where quality attributes serve as the basis for differentiation and market segmentation (Verhallen *et al.*, 2004).

Health is another personal need particularly because people tend to have different perceptions of health risks arising from consumption of particular foods (Aakkula *et al.*, 2005). Although every person needs nutritious food for a healthy life, health concerns arise from long-term side effects of eating certain types of foods. For example, increased publicity of the health hazards related with eating habits has enhanced demand for foods such as those with low cholesterol content (Angulo, Gil and Tamburo, 2005).

Once the basic and personal needs have been met, the next set of needs is sustainability and moral needs. While the broader definition of sustainable development rests on the PPP issues, considerations for sustainability may only become important issues for people when their basic and more personal needs have been fulfilled. This means that the importance of sustainability and environmental issues in the hierarchy of human needs is eroded when they cannot meet the lower level needs. However, consumers and other stakeholders increasingly demand sustainability and environmental protection and as a result, marketing firms increasingly respond to such demands in their corporate activities (see Van der Hoff and Roozen, 2001, in Dutch). Finally, when all of the foregoing needs are satisfied, people look for self-actualization, i.e., fulfilling ones desire to be what one wants to be.

In short, the fact that welfare, sustainability and quality appear at different levels of abstraction in the hierarchy of human needs has important implications about how they can be achieved, i.e., in the PPP context. The primary producers and the local communities first need food security before anything else. In order to enhance the profitability of their activities, primary producers must meet the consumer demands for different quality attributes. Hence, they may need to understand the quality attributes that consumers demand and determine how they can meet them. This means that small-scale producers not only need to have appropriate knowledge of quality demands but also the ability (e.g. having appropriate technologies) to produce and deliver the particular quality attributes. This study focuses on how small-scale primary producers can improve the quality of their products and at the same time implement practices that may protect NR. In view of the subjective nature of the food quality, the complexity of quality attributes and the inability of small-scale primary producers to meet consumer quality demands, this study considers basic quality attributes such as freshness. We focus on the ability of the primary producers to improve the quality of the fish.

1.5 Sustainability of the fisheries

Management of fisheries has generally been discussed within the framework of common property resources (CPR) (Demsetz, 1964; 1966; 1967; Hardin, 1968) because they are often not privately owned. Literature on NR develop concepts like maximum sustainable yield (MSY) which refer to the limits within which NR systems could be sustainably exploited.

Beyond such limits, the ultimate result is a tragedy characterized by reduced productivity, loss of biodiversity and reduced economic benefits that could be generated (Jul-Larsen *et al.*, 2003). In fisheries, MSY can be achieved by controlling fishing effort and methods by limiting the number, type and size of fishing gears; time and space (e.g., closed season and designated fishing zones) and number of fishermen (Jul-Larsen *et al.*, 2003).

Lack of private property rights over the fisheries subjects the fisheries to a "free for all" exploitation that renders sustainable management difficult and resources vulnerable to overexploitation (Hardin, 1968). The pursuit of self economic interests under free markets may imply that fishermen may continue fishing up until the fisheries are "commercially extinct"; i.e., until there are too little fish left to warrant profitable catching (FAO, 1995). As a result of the free market exploitation of the fisheries, the degradation of NR is alarming. Recent estimates on aquatic resources put 52% of fish stocks as fully exploited, i.e., fished at maximum biological capacity; 24% as over exploited, depleted or recovering from depletion and only 3% as being underexploited (www.msc.org/fish facts).

Some literature attributes the degradation of the fisheries and other NR to inappropriate institutional frameworks, enforcement costs and over-dependence on NR for survival especially in developing economies (Wade, 2004; Agrawal and Chhatre, 2006). On the one hand, as Holling (2000) notes, institutional frameworks driven by economic and industrial interests alone ignore that nature once disrupted may not be replaced with human engineering. On the other hand, institutional frameworks driven by conservation interests alone ignore the importance of an adaptive approach to economic development. For example, poor people who entirely depend on the fisheries may not guarantee sustainability if their daily basic needs are not satisfied.

In summary, sustainability of NR can be guaranteed if they are utilised within their biological carrying capacity. Whereas defining the measures for achieving MSY is one thing, persuading the resource users to implement them is another. Because free markets do not guarantee sustainability of the fisheries, alternative mechanisms may be needed. This study investigates if contracts can stimulate sustainable and quality-enhancing practices in developing economies as we elaborate in the next section.

1.6 Study approach and research questions

To-date, a number of procedures and standards to ensure food quality and safety have been developed such as the Codex Alimentarius (www.codexalimentarius.net), Hazard Analysis of Critical Control Points (HACCP) and Eurep-Gap (www.eurepgap.org). Similarly, a number of codes of conduct for protecting NR have been developed (e.g., FAO Code of Conduct for Responsible Fisheries; Convention of Biological Diversity, Marine Stewardship Council). These standards and codes of conduct for the sustainable use of NR, food safety and quality, constitute part of the global marketing and institutional environment within which small-scale actors must fit/operate. These standards and codes of conduct have generally been viewed as a promising road to sustainable development (see for example, the World Development Report, 2003) (World Bank, 2003). However, their implementation faces challenges across the globe. For example, the implementation and enforcement of the food quality and safety standards requires relatively huge financial and organizational resources that most developing economies do not often meet (Henson, Brouder and Mitullah, 2000; Henson and Traill, 1993). Consequently, the imposition of these standards and codes of conduct on developing economies is viewed as more of a trade barrier and constraint for the competitiveness of developing economies than for ensuring consumer safety (Hariss-White, 1999; Otsuki, Wilson and Sewadeh, 2001; Henson and Loader, 2001). These barriers challenge the extent to which they promise the road to sustainable development among small-scale primary producers in developing economies.

Therefore, alternative approaches are needed that would enable small-scale primary producers in developing nations to be part of international channels for their own welfare and without compromising the sustainability of NR. Crucial in this process is the responsible fishing by the implementation of sustainable and quality-enhancing practices. This study considers channel members' responsibility for channel level sustainability and particularly the use of contracts to stimulate channel members to implement sustainable and quality-enhancing practices. Contracts represent promises or obligations to perform particular actions in the future (Poppo and Zenger, 2002). Contracts detail promises, roles and responsibilities to be performed, specify procedures for monitoring and penalties for non-compliance, processes for dispute resolution and, most importantly, determine outcomes or outputs to be delivered.

The use of contracts has drawn arguments and counter-arguments (e.g., Williamson, 1985; Fafchamps, 2004). However, preference for contracts shows no signs of abating (Lusch and Brown, 1996). Their effectiveness depends on the institutional environment in which- and how- they are used (Cannon, Achrol and Gundlach, 2000). A number of factors influence the choice of contracts such as asset specificity, opportunism, uncertainties, social relations, among others which have been founded on the transaction cost economics and agency theory (Williamson, 1985), and social - and network theory (Wrong, 1968). Economics literature acknowledges that contracts minimise market failures and therefore enhance integration of small-scale primary producers into global market (Masakure and Henson, 2005; Key and Runsten, 1999; Singh, 2002). Further, given that fisheries are common property resources; this study also builds on property rights theory (Demsetz, 1967) to understand how and why channel partners may take different decisions regarding the sustainability of the fisheries.

The central question therefore, is how channel members collectively can take joint responsibility for a good balance of the people, profit and planet issues in the chain. More specifically, it focuses on whether contracts might be used to establish a better balance of the PPP issues in international marketing channels and what should be the terms of contract and governance structures that may stimulate welfare, profitability and sustainability at the primary production and in the channel as a whole. The thesis will address the following questions:

- What are the market failures that constrain the ability of primary producers to establish a better balance of the PPP issues in their activities?
- What terms of contracts and governance mechanisms can stimulate a better balance of welfare, sustainability and quality at primary production level?
- In what way can downstream channel members and other stakeholders help to stimulate welfare, sustainability and quality in the channel and especially at the primary production level and what would motivate them to do so?

1.7 Summary of the chapter

This chapter sets an argument that poverty and inequality in developing economies and degradation of NR continue to be widespread in the era when increased globalisation and integration of world economies offer opportunities for sustainable economic development. The chapter notes that the contributions of marketing to economic development have not fairly diffused across all segments of the society. The chapter also echoes the optimism that marketing can still change the world, particularly so for the developing economies. However, the chapter articulates that change can come if, among other things, small-scale actors who constitute the bulk of primary producers and custodians of the NR in developing economies are not only effectively integrated into the local economies but also in the global markets. The chapter discusses food quality and sustainability within the hierarchy of human needs. Finally, the chapter justifies the need for alternative governance regimes for sustainability and quality on the basis of the failure of free markets to guarantee sustainability of NR and the inability of developing nations to implement existing codes of conduct for sustainability and standards for food safety and quality due to resource limitations.

1.8 Structure of the thesis

The thesis is organized around six chapters (see Figure 1.3). Chapter (2) reviews relevant theories for channel governance. The chapter also discusses how contracts help to solve some of the market failures that small-scale primary producers face and then infers how contracts can help primary producers to implement sustainable and quality-enhancing practices.

Chapter 3 undertakes a situational analysis of the upstream part of the Nile perch channel to understand the context in which primary producers operate. The chapter identifies major market failure regarding welfare, quality of the fish and sustainability of the fisheries. Chapter 4 translates the theoretical framework from Chapter 2 to the empirical setting introduced in Chapter 3 and designs and tests fishermen's preference for sustainability and quality-enhancing contracts. These contracts oblige fishermen to implement sustainable and quality-enhancing practices. In return the contracts offer fishermen a number of terms of contracts intended to address some of the market failure they face.

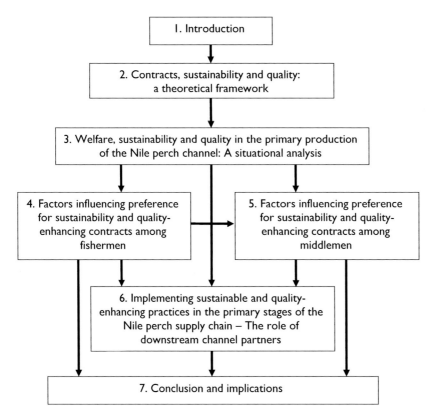

Figure 1.3. Structure of the thesis.

Chapter 5 applies the theoretical framework from Chapter 2 to the middlemen's setting also partly introduced in Chapter 3. The chapter also designs and tests middlemen's preference for sustainability and quality-enhancing contracts. The contracts also oblige them to implement sustainable and quality-enhancing practices. Middlemen are also offered a number of terms of contracts intended to motivate them. The chapter considers the double loyalty of the middlemen, i.e., to the buyers and to suppliers.

Chapter 6 takes the analysis to the downstream firms, i.e., processors, importers and retailers. This chapter brings the upstream problems to the downstream partners to determine why and how they may take responsibility and action to solve the upstream welfare, quality and sustainability problems. Finally, Chapter 7 synthesizes the results and discusses the policy implications for establishing a better balance of the PPP issues in the channel.

Chapter 2

Contracts, sustainability and quality: a theoretical framework

2.1 Introduction

In the era of market liberalization, globalization and expanding international agribusinesses, one major concern is how to integrate small-scale primary producers and entrepreneurs from developing countries into the global market economy. As relatively large organizational and financial resources become necessary for meeting the standards for participation in the global markets, small-scale primary producers and entrepreneurs in developing economies tend to lack both reliable and efficient production technologies, and guaranteed access to profitable input and output markets. In agriculture, contracts with large companies provide mechanisms for providing reliable and efficient production technologies and incorporating small-scale producers into global marketing channels (Glover, 1987; Key and Runsten, 1999; Masakure and Henson, 2005), despite arguments that contracts may also bring negative consequences (FAO, 2001; Singh, 2002). For instance, some literature argues that contracts result in more exclusion and marginalization of other poor primary producers who may not be able to secure contracts with large marketing firms (Porter and Phillips-Howard, 1997; Grosh, 1994). Learning from - and building on- the successes and problems of contracts in agriculture, and marketing and economics literature, this chapter develops a theoretical framework about contracts as a governance structure that incorporates small-scale primary producers into global market and most importantly, in a manner that enables them to implement sustainable and quality-enhancing practices.

This chapter is organized as follows. In section 2.2, we review some of the theories of channel governance and illustrate their strengths and weaknesses. The weaknesses of the current theories are further elaborated in section 2.3 that highlights market failures that developing economies continue to experience. Section 2.4 analyses some of the institutional frameworks that deal with degradation of natural resources (NR) and ensure food quality in international marketing channels. Section 2.5 introduces contracts as a response to market failures - illustrating some of the market failures that contracts (can) solve. Section 2.6 examines the problems that may arise from contracts while section 2.7 infers from economics and marketing literature about how contracts may stimulate sustainable use of NR and it is followed by section 2.8 that outlines the prerequisites for successful contracts. Section 2.9 integrates the preceding sections into a conceptual framework before section 2.10 draws conclusions for the chapter.

2.2 A review of theoretical approaches to channel governance

There are a number of approaches to channel governance such as perfect competition, the new institutional economics, social and network theory, property rights among others. Economic transactions take place under different forms of markets such as perfect competition, oligopolies, monopolies and monopolistic competition among others (see Henderson and Quandt, 1985). Among these markets, economic theory assumes that perfect competition gives the highest results in terms of allocating resources for the welfare of the society. The major assumptions of perfect competition are that, among others things: resources are perfectly mobile; products and demand are homogeneous (e.g., no quality differentiation); market information is symmetrically distributed; market access is without barriers; and supplier and buyers respectively seek to maximize profit and utility (see Hunt and Lambe, 2000 for a review). In practice, however, a market place consists of heterogeneous consumer demands that necessitate product differentiation (for example monopolistic competition); power imbalances and information asymmetries that affect market negotiations. These market place realities render the assumption that perfect competition gives the highest welfare for the society unrealistic and questions the efficiency with which price coordinates economic transactions (Van der Laan *et al.*, ed. 2000). Hence, firms shift to other mechanisms such as hybrid or plural arrangements to structure their transactions and circumvent the inefficiencies of perfect competition.

The new institutional economics through its branches of the transaction cost economics (TCE) and agency theory offers a different institutional arrangement to address the weaknesses of neoclassical economic theory of perfect competition (Coase, 1937; Williamson, 1985). The TCE envisages that contracts are designed to match known exchange hazards created by market failures such as specialized investments, opportunism, information asymmetry and uncertainty (Williamson, 1985). Agency theory focuses on economic incentives to minimise opportunistic behaviour of contracting partners (Eisenhardt, 1989b). The general proposition of both the TCE and agency theory is that transaction partners align the governance features to minimise exchange hazards and given a choice, they would choose institutional arrangements that minimise costs arising from mitigating exchange hazards (Coase, 1937; Williamson, 1985). However, the emphasis on minimising transaction costs undermines the possibility that other factors could influence the choice of governance structure (Rindfleisch and Heide, 1997).

Social - and network theory exemplify that the relational context within which economic transactions take place is critical for their performance. It has been demonstrated that social relations provide permissible limits of behaviour that enhance flexibility, solidarity, mutuality and information sharing necessary to adapt to turbulent market environments (Cannon, Achrol and Gundlach, 2000; Achrol and Kotler, 1999). Consequently, organizations strive to develop intimate social relations with customers and suppliers in order to mitigate against exchange hazards (Kim, 1999). Increasingly, social norms such as trust, commitment and network ties have become vital resources and antecedents for economic performance as demonstrated by the sheer volume of marketing literature (e.g., Atuahene-Gima and Li, 2002; Granovetter, 1973; 1985; Hewett and Bearden, 2001; Rindfleisch and Moorman, 2001; Wathne and Heide, 2004). For example, social relations mitigate against opportunism (Brown, Dev and Lee, 2000; Achrol and Gundlach, 1999) and enhance loyalty, shared values and respect (Brashear *et al.*, 2003; Sirdeshmukh, Singh and Sabol, 2002). However, social relations can be vulnerable to opportunism (Wuyts and Geyskens, 2005).

Governance for common property resources (CPR) such as the fisheries has largely been discussed within the framework of property rights theory. Property rights theory (Demsetz, 1967) exemplify the importance of the rights to use or generate income from a resource system. Property rights are the institutions that define or delimit the privileges granted to an actor or group of actors to use a resource system (Demsetz, 1964, 1967; Kim and Mahoney, 2002). Property rights theory deals with the processes by which the right of access and use of the resources are established. Such processes as open access, common or private determine the boundaries of the firms. The ability to appropriate rents from a resource system depends on how much property rights one holds over the resource and how they are protected (Foss and Foss, 2005; Kim and Mahoney, 2005). As protecting property rights incurs transactions costs, open and/or common property rights are seen as a deterrent to sustainable NR management because resource users do not often protect or invest in management of the resources from which they cannot guarantee the rents (Hardin, 1968).

In short, different theories approach economic problems from different angles and have their own strengths and weaknesses (see Table 2.1 for a summary). Relying on one theory may not adequately address multifaceted market failures such as degradation of NR resulting from - and leading to - poverty among rural communities in developing countries. Hence, as plural strategies are needed for governing complex transactions (Heide, 2003), integrated theoretical approaches to multifaceted market failures may also be needed.

2.3 Applying the theories of channel governance to developing economies

The debate on how to alleviate poverty and improve economic development in developing economies focuses on technological and institutional innovation and creation of favourable macroeconomic policy framework. The past decades witnessed a changing macroeconomic policy environment along with increased flow of trade, capital, technology and information across national borders especially from developed to developing economies in order to create a conducive environment to foster economic growth. However, for a number of reasons, some of which we seek to discuss in this section, the impact of such policy changes on economic development has not been substantial particularly in developing nations.

Table 2.1. Fundamental characteristics of the theoretical approaches to channel governance.

Theoretical approach	Theoretical perspective	Related governance mechanism	Focal decisions	Strengths	Weaknesses	Important references
Neo-classic economics	Perfect competition	Spot market negotiation	Price discovery	- Insights for profit maximization - Efficiency seeking mechanisms	Failure to recognise existence of information asymmetry and power imbalance in real life situation	Henderson and Quandt, 1985; vanTilburg et al., eds, 2000; van der Laan et al., eds. 2000; Ellis, 1988
Transaction cost economics	Transaction costs due to information asymmetry, asset specificity, opportunism and bounded rationality	- Vertical integration - Contracts	How to mitigate risks and uncertainties	Minimise transaction costs	Over-emphasis on costs ignoring any other factors in choosing governance structure	Williamson, 1985; Coase, 1937; Buvik and John, 2000; Rindfleisch and Heide, 1997
Agency theory	Importance of information, negotiation, monitoring and enforcement costs in determining terms of contract	Incentives (contracts)	What incentives can minimise opportunistic behaviour?	Insights into how to create incentives to minimise opportunism	Over-emphasis on costs silent on other factors such as value, corporate social responsibility, social relationship	Williamson (1985); Einsenhardt, (1989b); Dorward, (2001)
Social and network theory	Flexibility, solidarity, mutuality and information sharing in economic exchange	Trust and commitment	What norms should be promoted to enhance performance	Insights into flexibility and adaptability of relationships to risks and uncertainties	- Takes time, and effort to establish - Vulnerable to opportunism	Geyskens et al., (1998); Granovetter, 1973, 1985; Ganesan and Hess, 1997; Wrong, 1968
Property rights	Comparative assessment of incentives and rent seeking, constrained optimization	Institutions and Contracts	Property rights allocations, negotiation	Stakeholder view on property rights allocation	Difficult to enforce property rights, vested interests	Demsetz, 1967, 1966, 1964; Hardin 1968; Kim and Mahoney, 2002, 2005

One of the major policy reforms undertaken in most developing economies over the past few decades has been the structural adjustment programs that sought to bring about market reforms such as market liberalization, reduce state monopolies, eliminate input subsidies and price support to staple food crops among others (van Tilburg, Moll and Kuyvenhoven, eds. 2000; van der Laan, Dijkstra and Tilburg, eds. 2000). These reforms favoured the notion of free markets founded on the assumptions of the neoclassical economic theory of perfect markets. However, the market imperfections that ought to have been a result of state intervention in markets have not been completely solved (Jayne *et al.*, 2002; Southgate *et al.*, 2000). That is why success stories of efficient markets following the market reforms in developing economies are rare. The suboptimal performance of the free market phenomenon reflects persistent imperfections in market information, policies and institutions designed to enhance efficiency (Ramamurti, 1999; Kaplinsky and Morris, 1999). Moreover, free markets encourage pursuit of self-interests such as maximization of profit which eventually lead to over-exploitation of NR. Hence, following Coase (1937), we argue that the neoclassical economic theory may not adequately reflect the realities facing small-scale primary producers in developing economies and the context in which NR are utilized.

To that extent, the new institutional economics theory offers an alternative perspective to address some of the market failures developing nations continue to face such as information asymmetries, price risks, transaction costs and lack of access to modern technologies among others. Any efficient production process requires that producers have information about the needs of the markets in order to decide what, when and how to produce and where to sell. With more demanding markets, producers need information about market demands, for example, on the quality attributes. Small-scale primary producers in developing nations usually do not easily access such information because they are not only out of reach to information technologies and media (e.g. television, telephones, fax, internet, newspapers), but also geographically scattered, uncoordinated and mostly, illiterate. Lack of market information reduces their ability to bargain for better compensation for their products. In the end, primary producers tend to be at the mercy of other market actors such as the middlemen who tend to be well informed about markets (Ellis, 1988). Because small-scale producers in most developing countries are scattered in rural areas, they often have no access to better transport and market infrastructure which tend to be localised in towns and cities. Hence, coupled with lack of market information, they spend more time, effort and resources in search of input or output markets thereby increasing transaction costs.

Modern production technologies that are needed to produce for modern markets are generally more costly than traditional technologies that small-scale primary producers in developing nations can afford. Yet, the traditional technologies do not often meet the modern production standards. As a result, most producers require credit facilities to finance their production technologies. But the structure of credit markets does not favour poor small-scale primary producers who often have no collateral (Coleman; 2006). Although there are some micro-finance institutions to reach out to small-scale primary producers, they too must demonstrate economic viability of their existence (Rahman, 1999; Basu, 1997). Consequently, such institutions do not always reach out to the poor who need credit the most and yet their level of operation may not contribute to the economic viability of the credit institutions. Inevitably, efficient and equitable development outcomes may depend, in part, on whether formal financial institutions continue to leave low-wealth producers tightly constrained in their access to credit for production facilities and also whether market channels are governed in a way that reward members according to their contributions to market offerings.

In short, as market place continues to change, more imperfections may emerge and the more free markets may fail to efficiently allocate resources. The high information asymmetries, lack of access to credit and high transaction costs observed in many developing countries may explain why small-scale primary producers tend to be excluded and marginalized from participating in global markets. Unfortunately, the macro-economic market and policy reforms intended to minimise such market failures have not achieved their goals (Southgate *et al.*, 2000). The sustained market imperfections have negative impact on the livelihoods and welfare of small-scale producers – the bulk of producers in developing nations. Ultimately, NR such as fisheries and forestry on which the poor depend for survival will continue to be over-exploited and degraded. Next we discuss how some of international marketing channels try to circumvent these market failures.

2.4 Dealing with market failures in international marketing channels

A number of institutions, institutional frameworks and processes that constitute an institutional environment in which marketing channels operate evolve as market failures emerge. Economics literature defines institutions as rules of the game, i.e., devised constraints that shape human interaction through structuring incentives whether social, economic or political (North, 1990). Marketing literature refers to institutions as governance structures or institutional frameworks used to establish and structure the exchange relationships (Heide, 1994; Buvik and John, 2000). These can be rules or institutions that simultaneously empower and constrain actors to pursue activities consistent with predefined patterns of acceptable channel behaviour (Grewal and Dharwadkar, 2002). Whether one inclines to economics or marketing literature, institutional frameworks or governance structures arise in response to pressures and conditions of a particular era or context (Grewal and Dharwadkar, 2002). For example, economic institutions emerge to perform economic activities that cannot be guaranteed by spot markets (Ellis, 1988). We use institutional frameworks or governance structures interchangeably referring to mechanisms that are used to coordinate activities between transaction partners.

International agribusiness channels use different institutional frameworks to deal with issues that spot markets cannot guarantee such as food quality and protection of NR. Marketing channels impose or motivate members to adopt certain standards (Grewal and Dharwadkar,

2002). *Impositions* involve use of regulatory powers that make compliance with the standards mandatory. This is often the case when the activities of marketing channel are in conflict with societal interests such as public health. The imposition of food safety standards by major export regions such as the EU is one such example. However, not all impositions tend to be regulatory. For example, marketing channels impose non-regulatory standards on their members to win confidence of their customers. One such example is the imposition of the food safety and quality standards on the European retail chains that are certified with EurepGap (www.eurepgap.org). *Motivations* involve creating incentives for channel members to conform to the standards. The environmental certification institutions such as the Marine Stewardship Council offer incentives such as use of the eco-labels through which marketing channels may charge premium prices for their fisheries products (www.msc.org).

However, either impositions or motivations may not always be effective in developing economies where information asymmetries, poor access to production technologies, risks and uncertainties affecting primary producers are complex. Further, primary producers also have to deal with inefficiencies and ineffectiveness of the institutions intended to address such market failures (Southgate *et al.*, 2000; Ramamurti, 1999; Ng and Yeats, 1997). Hence, an alternative institutional environment, i.e., market institutions, institutional frameworks and processes are needed that can solve market failures in developing economies. We focus on contracts that, based on the lessons from economics and marketing literature, can minimise some of the market failures for primary producers (Key and Runsten, 1999; Masakure and Henson, 2005).

2.5 Contracts as a response to market failures

An understanding of why firms choose contracts over other governance arrangements such as spot markets is important to envisage the possible outcome of contracts in promoting sustainable and quality – enhancing practices among small-scale primary producers. This section builds on marketing, management and new institutional economics literature to illustrate motivating factors for small-scale producers to enter into contracts with buyers.

Contracts are agreements to undertake future transactions under predefined conditions (e.g. Poppo and Zenger, 2002). Contracts outline a selection of promises, obligations, outcomes, procedures for monitoring and dispute resolution, and penalties for non-compliance. Management and marketing literature often recognises two major types of contracts: legal contracts – those enforced by law, and relational contracts – those embedded within social norms (Cannon, Achrol and Gundlach, 2000). This literature builds on the logic of TCE that emphasize the need to minimise exchange hazards arising from asset specificity, information asymmetry, opportunism and uncertainty (Williamson, 1985).

Economics literature refers to contracts as forward agreements between producers and processing and/or marketing firms for the production and supply of products (Singh, 2002;

FAO, 2001; Grosh, 1994). This literature recognises three general types of contracts: market, resource-providing and management contracts (FAO, 2001; Key and Runsten, 1999; Singh, 2002). *Market- specifications contracts* are those in which producers and buyers agree on the terms and conditions for future sale and purchase of products. The major elements of market-specifications contracts include agreement on price, quality and delivery schedules (Key and Runsten, 1999, Singh, 2002). *Resource- providing* contracts are those in which buyers provide selected production inputs and technical services beside an agreement on buying and supplying of the products (FAO, 2001; Key and Runsten, 1999; Glover, 1984). *Management contracts* are those in which producers agree to follow recommended production specifications preferred by the buyer in return for both purchase of the products and provision of the production inputs and technologies (FAO, 2001; Key and Runsten, 1999; Singh, 2002). The choice of any of these types of contracts may depend on a number of factors such as type of product, legal framework, market conditions and past experiences between transaction partners (Key and Runsten, 1999).

Contracts are considered as an institutional response to some of market imperfections for credit, information, technology transfer and access to input or product markets (Grosh, 1994; Key and Runsten, 1999). Factors influencing transaction partners to engage in contractual arrangements have been documented (Williamson, 1985; Houston and Johnson, 2000; Grosh, 1994). Small-scale primary producers engage in contracts depending on the market failures they face and their vulnerability to risks (Masakure and Henson, 2005). Literature suggests a number of market failures that small-scale producers face for which contracts are suitable alternative structures such as (1) access to credit for production inputs, technologies and services, (2) price risks and information asymmetries, (3) access to profitable output markets, and (4) product quality (Key and Runsten, 1999; Masakure and Henson, 2005; FAO, 2001; Singh, 2002; Glover 1984, 1987).

2.5.1 Access to credit for production inputs, technologies and services

Access to credit is important for small-scale producers who often cannot afford modern technologies for efficient production. As the structure of the formal credit markets is often not suitable for small-scale primary producers, agro-business firms act as lenders to their contracted producers (Key and Runsten, 1999; FAO, 2001). With such credit arrangements, the buyer, on the one hand, is assured not only of product supply but also that the credit is spent on the intended production. The loans are often distributed in kind and supervised by the lenders themselves. On the other hand, the producers are assured of credit which they may not easily obtain from alternative sources and also markets for their products (Masakure and Henson, 2005; FAO, 2001).

New production technologies are required to enhance productivity as well as to ensure that the commodity meets market demands such as quality standards. Markets for inputs or services needed for efficient production are thin or missing in developing countries (Key and Runsten, 1999). For example, in Africa, agriculture services provided by public institutions have been disrupted by structural adjustment measures, leaving the private sector having yet to fill the void created that adversely affect input availability and use. Small-scale producers involved in contractual production tend to be supported in terms of production inputs and extension services (Grosh, 1994; FAO, 2001). In that way, the producers may have a better understanding and use of husbandry practices necessary to achieve intended production standards and quality. In the end, small-scale producers engaged in contract production tend to be in better a position in accessing production inputs, technologies and services than their counterparts outside contractual arrangements.

2.5.2 Price risks and information asymmetries

The returns suppliers receive on the open market depend on the prevailing market prices as well as on their ability to negotiate with buyers. This can create considerable uncertainty which, to a certain extent, contracts overcome. Often, buyers indicate in advance the price(s) to be paid and these are specified in the agreement thereby minimising price risks for producers (Masakure and Henson, 2005; Grosh, 1994). Further, when production is information intensive and producers lack resources to acquire appropriate information, contracts tend to be the main source of information for market demands.

2.5.3 Product markets

Small-scale producers are often constrained in access to profitable markets for their products either due to lack of knowledge about the markets or distance to the markets. Consequently they sell their products in local markets that are often less competitive. Contracts with buyers offer market guarantees for small-scale producers (Masakure and Henson, 2005). Even where other market outlets exist for the same crops, contracts may offer significant advantages to producers because they reduce searching and negotiation costs. Contracts may also minimise transport costs as buyers may offer transport (Masakure and Henson, 2005; FAO, 2001).

2.5.4 Product quality

Quality is one of the critical demands for modern markets. Markets for fresh and processed products, for example, require consistent quality standards and management systems. Modern food markets increasingly move to a point whereby suppliers must conform to regulatory and non-regulatory controls, for example, regarding food safety, traceability and product labelling. These require that suppliers need to be confident of the source of their products in the event that food safety problems arise. Besides, with close supervision by the buyers, access to production facilities (such quality management facilities) and information on quality standards, producers may be committed to meet quality standards because they are bound to loose if their product does not meet quality standards as agreed in the contract (Key and Runsten, 1999; Masakure and Henson, 2005). Also contractual arrangements make quality assessment more manageable. Although product quality from small-scale primary producers may be difficult to guarantee with very high certainty, it may certainly be better off to obtain from contracted producers who have stronger motivation to produce good quality products than sourcing the same products from spot markets (Grosh, 1994; FAO, 2001).

2.6 Problems with contracts

The fact that contracts can solve some of the market failures that small-scale primary producers face does not necessarily imply that they cannot create problems. One major problem that has been raised in literature is the effectiveness of contracts to coordinate transactions under situations of high uncertainty (Williamson, 1985). In order to circumvent this, firms choose other mechanisms or use contracts in combination with other mechanisms depending on the context of the transactions. For example, Houston and Johnson (2000) suggest that actors are likely to choose joint ventures over simple contracts when the degree of asset specificity is high. Heide (2003) suggests that under high information asymmetry, firms are likely to shift from exclusive reliance on market contracting to plural governance, i.e., contracting and vertical integration. Hence, although TCE argues that information asymmetry, specific assets, opportunism and uncertainty necessitate contracting, they may also limit the extent to which actors can rely on contracts to deal with these market failures. Economics literature also argues that contracts between buyers and small-scale producers can create negative consequences for both small-scale producers and buyers (Grosh, 1994; Porter and Phillip-Howard, 1997; Fafchamps, 1996; 2004; Singh, 2002). We briefly discuss some of the problems for producers as well as the buyers.

2.6.1 Problems for small-scale producers

Literature suggests a number of problems that small-scale primary producers may face when they engage in contracts with buyers such as, (1) increased risks; (2) manipulation; (3) domination and (4) indebtedness, among others (Grosh, 1994; Glover, 1984, 1987).

Increased risks: Small-scale producers venturing into new contractual arrangements may have to be prepared to balance their expected returns with a possibility of greater risks. Risks may arise from production failures due to factors beyond the producers control such as weather or ecological factors (Grosh, 1994). Risks may also arise when the buyer faces unexpected (market) uncertainties which may ultimately affect his commitment to the terms of contract (FAO, 2001).

Manipulation: Small-scale producers may be manipulated by buyers. Although this type of problem may be difficult to discover, there are many ways in which producers might be manipulated. For example, buyers can manipulate quality standards to control deliveries (Glover, 1984). If production exceeds the buyers' requirements, buyers can covertly raise quality standards in order to reject excess production without appearing to breach the contract (FAO, 2001). Such practices may lead to confrontations which can be minimised if they have methods to resolve disputes over grading irregularities or forums where concerns and grievances relating to such issues can be resolved.

Domination: In cases where producers are in a weak bargaining position or do not fully understand the implications of the contract, the buyer can draw up an agreement which, when enforced, may easily manipulate the producer (Glover, 1987). Further, the monopoly over the purchase of product by the buyers can lock producers into the transaction without chance to try other buyers (Key and Runsten, 1999). Producers who might have invested into such relationships may be affected more because they may not easily exit the relationships.

Indebtedness: Although one of the major attractions of contractual arrangements for small-scale producers is the availability of credit facilities, they may face considerable indebtedness to the buyer. For example, when producers fail to produce as expected due to different reasons, they may defer the repayment of the production input loans which may lead them to be indebted to the buyer (Singh, 2002; FAO, 2001).

2.6.2 Problems for buyers

Problems of contractual supply for the buyers can be examined in light of the alternative mechanisms especially spot market purchases. The problems for buyers might include, (1) opportunism, (2) producer discontent, (3) enforcement, (4) supply unreliability, indebtedness and shared risk.

Opportunism (extra-contractual marketing and input diversion): Opportunism is one major justification and also a problem for contractual arrangements (Williamson, 1985). Opportunism entails actors acting out of self- rather than mutual- interest. Opportunistic producers may take advantage of information asymmetry and bounded rationality of the buyers to exploit any loopholes that might not have been adequately specified in the contracts. On the one hand, sale of products to other parties outside the contract may not be easily controlled when alternative market outlets for the product exist (FAO, 2001; Fafchamps, 1996, 2004; Grosh, 1994). This problem could be acute where alternative market outlets offer better prices than what contractual buyers offer. On the other hand, buying of products from non-contracted producers could also be a problem to buyers. This may occur when non-contracted producers take advantage of, for example, higher prices offered by an established buyer. Products from non-contracted producers may be filtered into the buying system by outside producers through contracted friends and family relatives. Whereas such practices may be good for buyers who face supply shortages from contracted producers, it may also be difficult for buyers to regulate production targets and other quality aspects when they face excess supply.

In contracts where production inputs and facilities are provided to producers, there is also a danger that inputs might be diverted. In farming, for example, farmers may be tempted to use inputs supplied under contract for unintended crops or even to sell them to obtain immediate cash (FAO, 2001). Such behaviour is bound to affect the contracted product supply quantity and quality. These problems may be overcome by improved monitoring, issuing of realistic quantities of inputs and also creating good social relations with the producers. In addition, as marketing literature suggests, the advantages arising to producers from access to resources, which they would not otherwise obtain without contracts, may minimise opportunistic intentions (Rokkan *et al.*, 2003; Ghosh and John, 1999),

Producers' discontent: Buyers may face problems of discontent from producers that may arise from a number of situations. For example, producers who experience discriminatory treatment, late payments, abrupt changes in terms of contracts such as prices and poor social relations with the buyers may generate dissent (FAO, 2001). These and other circumstances may cause hostility towards the buyer that may result in producers either withdrawing from projects or undertaking extra-contractual selling of their products. Ultimately, buyers may have problems such as supply gaps which contractual arrangements intended to avoid.

Enforcement: Compliance with the terms of contracts depends to some extent on whether or not enforcement mechanism exists that penalizes breach of contract (Fafchamps, 1996). Fafchamps (2004) analyses contract enforcement in developing economies particularly Africa. He outlines different mechanisms that are used to enforce compliance such as harassment and retaliation. Retaliation may also be inflicted by other people closer to contract parties. For instance, Antia and Frazier (2001) suggest that a principal may not use severe enforcement measures when a central agent violates a contract for fear of retaliation from other agents close to him. Such enforcement problems may leave the buyer in a precarious position to deal with contract violations (Grosh, 1994) and may compromise contractual standards.

Supply unreliability, indebtedness and shared risks: One of the possible problems for buyers may be the producers' failure to supply agreed products that may jeopardize future sales. Even with the best management contracts, buyers always run the risk that producers fail to honour agreements or fail to produce due to factors beyond their control such as weather or illness. However, dealing with contracted producers enables buyers to share the risk in the event that they agree to defer the producer's repayment of the loan (Grosh, 1994, FAO, 2001). In addition, the buyer may face risks arising from fluctuation in market conditions. For example, by agreeing to pay fixed prices to the producers, the buyer runs a risk of losing if output markets offer less than what the buyer promised to the producers. In that case, the buyer may also defer some payment to the producers thereby becoming indebted or risking producer discontent or extra-contractual selling of products.

In summary, contracts may create problems for both producers and buyers. However such problems can only be a deterrent to the use of contracts if they can be avoided by using alternative mechanisms such as spot markets. It is evident from literature that there are certain situations in which both buyers and small-scale producers may be better off having contractual arrangements than spot markets (Dorward, 2001; Key and Runsten, 1999; Williamson, 1985). But in spite of the extensive literature on how contracts (may) solve market failures for small-scale primary producers in agriculture, there is little or no literature to suggest that contracts (may) solve such market failures as degradation of NR - one of the major market failures levelled against the current marking systems. We discuss this next.

2.7 Contracts and sustainable practices in NR

Contracts as an alternative institutional framework for addressing market failures have largely been demonstrated in profit-oriented economic transactions. As such, the use of contracts for promoting sustainable practices in NR has not been the focus in available literature. This is against the background that depletion of the NR such as fisheries is one of the major market failures and challenges of the existing marketing systems (Wilkie and Moore, 1999; Aggarwal, 2006). Consequently, there is limited (or no) literature to suggest how and if contracts could be used to enhance sustainable practices particularly so among small-scale primary producers in developing economies. Nonetheless, we infer from the preceding discussions about how contracts could promote sustainable practices in NR.

The preceding discussion demonstrates that small-scale primary producers may appreciate contracts that minimise the risks and uncertainties that they face. Although the market failures that small-scale primary producers in developing economies face may be generalised, those in NR face additional and unique risks arising from degradation and depletion of NR. Declining production from NR such as the fisheries erodes the fountain of their livelihoods and survival. In addition, the nature of the property rights for NR, i.e., common property rights, makes it difficult for small-scale primary producers to protect NR they depend on from exploitation for private interests that has increased with globalization and market integration (Aggarwal, 2006). Hence, although contracts come along with other risks for small-scale primary producers (as discussed above), those in NR may face much more risks with the degradation of NR under, for example, spot markets than the risks they may face with contracts per se.

Like other small-scale primary producers in developing countries, those in NR face heightened risks given that they may have limited exit options to enter into other formal sectors of the economy due to limited levels of formal education. Hence in view of their high dependency on the NR, protecting the NR may be a motivation by itself for small-scale primary producers to appreciate sustainability-enhancing contracts. In order to make them effective, contracts may have to address both the general market failures such as price risks, information asymmetries, access to credit and those specific to NR such as appropriate technologies for sustainable production. This implies that having a better understanding of the context in which small-scale primary producers operate is critical for the design of realistic and appropriate contracts for promoting sustainability of NR. Moreover, as FAO (2001) notes, the success of contracts depends on the prerequisite conditions. We elaborate next.

2.8 Prerequisites for successful contracts

Contracts are necessary to address market failures but they also create problems. However the success of contracts, as Cannon, Achrol and Gundlach (2000) suggest depends on where and the conditions under which they are applied. FAO (2001) suggests that no contract should be ventured into unless some basic preconditions are evaluated. An assessment of the context in which contracts would be applied is important in order to design and evaluate the benefit that may accrue from them. Such prerequisites include, (1) profitable markets, (2) physical environment, (3) social environment, (4) regulatory environment and, (5) property rights.

2.8.1 Profitable markets

For any contractual arrangement to be successful there has to be a profitable market for both producers and buyers. If producers, buyers or both fail to achieve consistent and attractive economic gains or profits, a contract venture may collapse. Hence, the decision to engage in contractual supply may have to be done with the expectation that, subject to other conditions, it will be profitable (FAO, 2001). Guaranteed, regular and attractive incomes may encourage contracting partners to make long-term commitments.

2.8.2 Physical environment

One of the preconditions for investment, for example, in rural areas is the existence of communication facilities such as roads, transport, telephones and other telecommunication services and facilities. Reliable power and water supplies are vital for agro-processing and exporting of fresh produce such as fish. Availability of regular cargo airfreight is important for export of fresh products. These may affect the extent to which foreign buyers may want to engage in contracts with producers from developing countries or the ability of exporters in developing economies to secure supply contracts in major export markets where, for example, timeliness of delivery is an important parameter of competitive advantage.

2.8.3 Social environment

Social factors are also basic factors in establishing contractual relations. Many rural communities tend to be influenced by traditional practices in their ambitions, material needs or timeframe to achieve them. Cultural issues could be crucial with foreign investors as the cultural divergence with local producers could be high. Contracts might disrupt social relations within communities which may in turn affect the success of a contract venture (Singh, 2002). For example, if producers are chosen on the basis of resource endowment, contracts may widen pre-existing economic disparities and lead to animosity and resentment from those excluded from contracts (FAO, 2001).

Some literature argues that exclusive reliance on contracts per se may constrain actors in their ability to evaluate partner's performance and make correct evaluations of the performance problems if they occur (Heide, 2003). Further, contracts may lock up producers into relations for which better alternatives exist thereby limiting their chances to benefit from alternative markets. Marketing and other related literature suggest that social relations may be barriers to switching to new relations (Wathne, Biong and Heide, 2001; Fafchamps, 1996). In the same way, social relations may bar actors from engaging in contractual arrangements especially if they fear that their existing socio-cultural relations and networks may be disrupted (Singh, 2002). Hence understanding the social environment in which the contracts would be applied may help to design contracts that are sensitive to the social issues.

2.8.4 Regulatory environment

Public institutions may have a role in the success or failure of the contracts. For example, a relevant legal framework and an efficient legal system are preconditions for settling contractual disputes and also protecting national heritage and resources (Fafchamps, 2004). FAO (2001) defines this as a regulatory role for the government. In addition, government can play a developmental role by providing adequate public utilities and services such as market infrastructure, roads, power, water, and communication facilities necessary for the success of contracts.

2.8.5 Property rights

Property rights over the resource system may be important especially in NR where common property rights tend to be viewed as an important factor that limit the extent to which resource uses take management responsibilities (see Hardin, 1968). In common property resources, contracting only a section of resource users may not easily translate into the sustainability of the NR if other resources users outside the contracts continue with unsustainable practices. Furthermore, it may not be motivating for the small-scale producers to engage in sustainability-enhancing contracts if they cannot guarantee rewards for protecting the resources.

In summary, whereas contracts may solve some market failures, they may also create problems for contracting partners. Understanding the context in which contracts would be applied may be crucial for the design and implementation of appropriate and suitable contracts for enhancing the implementation of sustainable practices among primary producers.

2.9 Contracts, sustainability and quality - summary

Literature show that contracts as alternative governance structures overcome a number of market failures which small-scale primary producers face in developing countries. Contracts help primary producers to access production resources, market information and improve quality (Masakure and Henson, 2005; Key and Runsten, 1999) and also distribute risks between contracting partners (Grosh, 1994). More importantly, contracts may enhance the integration of small-scale primary producers into international marketing channels (Masakure and Henson, 2005; Glover, 1987; Key and Runsten, 1999). However, contracts may also create a number of problems for contracting parties (e.g., Fafchamps, 2004; 1996; Glover, 1984; 1987; Grosh, 1994; Key and Runsten, 1999; Porter and Phillips-Howard, 1997; Singh, 2002; Williamson, 1985). This chapter provides a preview of contracts as alternative institutional framework to deal with market failures that spot markets fail to solve. The chapter highlights the conditions under which contracts may or may not solve market failures. This chapter also infers that by minimising market failures, contracts may also promote sustainable practices especially among small-scale primary producers.

Contracts in the primary stages of international channels may ensure product quality from the source. This is crucial for fresh products where quality cannot be guaranteed anywhere else in the channel if not from the source. Similarly, protection of NR cannot be guaranteed in marketing channels if there are no incentives at the primary stages. Consequently, major improvements in quality and sustainability of NR in marketing channels can be expected if such improvements start at the primary stages. Contracts may provide a better framework to define the conditions for small-scale primary producers and entrepreneurs to implement sustainable and quality-enhancing practices. For example, management contracts that oblige producers to implement recommended production specifications in return for market guarantees and access to production resources may promise a potential framework for primary producers to implement sustainable and quality - enhancing practices.

However, as problems with contracts may not be ignored, one way to minimise potential problems with contracts is to understand the context in which contracts would be applied and design appropriate contracts that should minimise risks and uncertainties for both buyers and producers (FAO, 2001). In addition establishing the right forums for conflict resolutions would be necessary. Hence, the problems that producers and buyers may face with contractual arrangements may not necessarily be a black and white deterrent to try contracts in NR.

Therefore, the choice of contracts may depend on the market failures that need to be addressed within the context in which contract partners operate. The specific market failures that contract partners face may determine the specific terms of contracts and the context in which contract parties operate may influence the extent to which they may appreciate particular contracts (Figure 2.1).

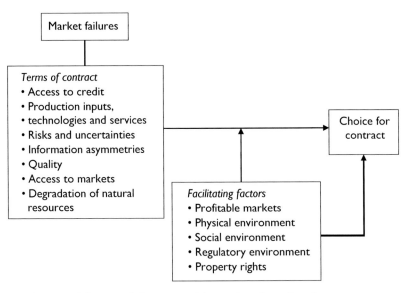

Figure 2.1. Conceptual framework for sustainability and quality-enhancing contracts.

As specific market failures and their importance may differ along the chain, this study focuses on, (1) primary producers (fishermen), (2) middlemen and (3) downstream companies, i.e., processors, importers and retailers and other stakeholders.

Primary producers: The study on primary producers is crucial because as primary custodians of the fisheries, fishermen also have private interests to pursue. On the one hand, their decisions on how - and how much - to produce may affect the sustainability of the fisheries. For example, if they use bad fishing gears or if they fish in breeding zones, it may lead to further degradation. Further, poor incentives to protect the fisheries for public interest coupled with pressing welfare needs may result, as is already evident, in resource over-exploitation and degradation. Moreover, free market mechanisms that support pursuit of self-interests, and globalization of world markets contribute to the degradation of local ecosystems (Aggarwal, 2006). In order to devise meaningful contracts, we need to understand the specific market failures fishermen face and the environment in which they operate.

Middlemen: Analysis of the subsequent channel stages is equally important. To begin with, production and marketing sectors are not mutually exclusive. Marketing sector is assumed to drive the production strategies which in turn may impact on the marketing strategies. Middlemen who have double loyalty, i.e., to suppliers (fishermen) and to buyers (downstream companies) may contribute to how primary producers undertake their activities. It is thus important to investigate how and if middlemen can implement sustainable and quality-enhancing practices in their buying and supplying practices.

Downstream companies: Downstream channel partners in modern markets tend to influence the way upstream channels operate for example, through imposing quality standards. As sustainability of NR is not only becoming an important market demand, but also a corporate social responsibility issue for which downstream channels are being held responsible (Bansal and Roth, 2000), it is important to investigate why and how downstream companies can promote sustainable practices in the upstream part of the channel. Such an analysis should focus on how downstream channel partners can help to minimise the market failures that primary producers face in a manner that improves their welfare and also motivates them to implement sustainable and quality-enhancing practices.

2.10 Conclusion

In conclusion, whereas implementing sustainable and quality-enhancing practices is important for any channel, achieving it a manner that relieves small-scale primary producers of the market failures they face would in itself be an important contribution to sustainable development. Contracts that may help to achieve that may ultimately be an integral part of the governance structures not only for integrating small-scale primary producers into international marketing channels but also for improving the welfare of the local communities - an important pillar of sustainable development. Since understanding the prevailing conditions surrounding the actors is a prerequisite for the success of contracts, the next chapter (3) undertakes a situational analysis of the conditions under which fishermen undertake their activities. The chapter seeks to identify the market failures and how fishermen cope with them. Subsequently, Chapters 4 and 5 design and test different contracts for fishermen and middlemen's preference for the contracts in order to determine factors that may motivate them to implement sustainable and quality-enhancing practices. Finally, Chapter 6 investigates whether and how downstream partners can help to promote welfare, sustainable and quality - enhancing practices in the upstream part of the channel.

Chapter 3

Welfare, sustainability and quality in the primary production stages of the Nile perch channel: A situational analysis

3.1 Introduction

This chapter undertakes a situational analysis of the context in which fishermen and middlemen operate. The chapter seeks to address, (1) what market failures fishermen face, (2) how they cope with them and, (3) how they could be addressed. The chapter provides a research setting in which preference for contracts that oblige fishermen and middlemen to implement sustainable and quality- enhancing practices is subsequently tested. The chapter is organized as follows: Section 3.2 outlines conceptual framework for the study followed by section 3.3 that outlines the research methodology. The results and discussion follow in section 3.4. The implications of the results are given in section 3.5. Study limitations in section 3.6 conclude the chapter.

3.2 Conceptual framework

In order to identify market failures that fishermen and middlemen face and how they cope with them, an exploratory case study is undertaken. Case study research is one way of undertaking social research often suitable and preferred when (1) "how" and "why" questions are being asked; (2) researchers have no or little control over the events; (3) the focus of research is on contemporary issues within a real life situation and (4), when dealing with relatively new research topics (Eisenhardt, 1989a; Yin, 2003). Case studies may accomplish different outcomes such as to provide a description, test or generate hypotheses (or theory) (Yin, 2003; Eisenhardt, 1989a). Since this chapter focuses on contemporary channel activities for which we do not have control, it befits to undertake a case study

In this case study, we consider a number of concepts. First, we analyse the structure of the channel. According to Coughlan *et al.*, (2001), *channel structure* refers to the configuration of the channel stages and actors in the different stages. Analysing channel structure will enable us to understand channel relations especially power and dependency that may have implications on how actors negotiate for the terms of transactions (Heide and John, 1988). According to Coughlan *et al.* (2001), "*Power* is the ability of one channel member to get another member to do something he otherwise would not have done." Power arises from structural position of the actor in the channel (Heide and John, 1992) or ownership of and/or access to resources uniquely important to undertake particular transactions (Argyres and Liebeskind, 1999). Imbalances in power enable powerful actors to dictate and control the activities of the less powerful and therefore *dependent* actors (Gundlach and Cadotte, 1994). We will examine the sources of power and how power affects transaction relationships.

We will also analyse governance mechanisms. *Governance* mechanisms refer to the tools that are used to coordinate activities between transaction partners (Hendrikse, 2004). There are different mechanisms that are used to coordinate economic transactions such as contracts, hierarchies or spot markets or a combination of these (hybrid governance) (Heide, 2003; Williamson, 1985). Actors choose different governance mechanisms depending on the pressures and/or incentives that they encounter. This chapter analyses how transactions in the upstream are governed.

This chapter assesses quality, sustainability and food security for local communities. At primary level where no major value adding activities are undertaken, *quality* refers to the basic attribute of freshness. An estimate of freshness can be obtained by defining criteria related to changes in sensory attributes like appearance, odour, colour and texture, which can be measured by sensory or instrumental methods (Olafsdottir *et al.*, 2004). Furthermore, handling, processing and storage techniques, time and temperature may also affect freshness. Improving quality requires special facilities such as cooling facilities. This chapter assesses how primary producers manage fish quality with or without specialised facilities.

Sustainability of the fisheries is, in this study, considered from the implementation of sustainable practices such as use of recommended fishing gears. Using recommended fishing gears promotes sustainability because they protect the juvenile fish to remain in water and grow (Jul-Larsen *et al.*, 2003). In order to achieve sustainability, fishermen are expected to use recommended fishing gears and middlemen are expected to buy recommended fish sizes. Further, public institutions are mandated to ensure that sustainable practices are enforced. This chapter therefore examines how different players undertake their activities and responsibilities with respect to sustainability.

One of the basic welfare needs of the local communities is *food security*. Access to fish is one way of ensuring food security for local communities. This chapter will examine how local communities access Nile perch fish for *food security*.

3.3 Research methodology

3.3.1 Selection of cases

In order to select cases, we followed procedures outlined by Eisenhardt, (1989a) and Yin, (2003). The Nile perch fish supply chains originate from designated landing sites or beaches where fishermen and middlemen transact. As random sampling is not essential in a case study (Eisenhardt, 1989a), we selected beaches based on ease of access and landing of Nile perch. A list of designated landing sites was obtained from Kenya Marine and Fisheries Research Institute (KMFRI) - a semi-autonomous research institute mandated to undertake marine and fisheries research in Kenya. We conducted preliminary visits and interviews in 15 beaches. We noted that there were more similarities than differences with respect to channel organisation, fish handling facilities, fish buyers and other transaction activities. We finally chose eight landing sites that were landing significant amount of Nile perch and also with good patronage by middlemen. Good patronage by the middlemen was important for ease of getting them involved in the study.

3.3.2 Respondents

Key informants were members of the beach management units (BMU). The BMUs are local administrative committees comprising fishermen and local traditional leaders. The BMUs manage the activities of the beaches including monitoring and ensuring sustainable fishing practices, hygiene standards, marketing, security and conflict resolution, among others. In addition to the key informants, other fishermen and middlemen were also interviewed. Fishermen had fishing and fish trading experiences ranging from 2 years to over 40 years while middlemen had trading experience ranging from less than a year to over 5 years. This enabled us to obtain a wide retrospective account of the fishing and fish trading activities. Some fishermen and middlemen had changed roles, for instance, from being fishermen to middlemen or fishing and trading other fish species to Nile perch.

Although our focus was on the landing sites where fishermen and middlemen are the integral players, the upstream channel ends up at the processing factories. Hence to have a complete picture of the structure of the upstream channel and also for verification of some of the issues raised at landing sites, three processing factories were also interviewed. Some of the fishermen, middlemen and processors were transacting with each other which made verification of some issues easier from a channel perspective.

3.3.3 Data collection

Data were collected between February 2004 and March 2005 through literature review, focus group discussions, individual interviews and personal observations consistent with Yin (2003) and Stewart and Shamdasani (1990). One research assistant helped in data collection. Although the research assistant was quite experienced in conducting interviews, he was trained in the specific issues of this study. His main task was to help transcribe the interviews and also to translate the interviews into local language where necessary.

Literature review: While some literature was sought from the internet, most of it was obtained from the KMFRI library. Given its strategic position in the sector, the institute has literature on varying aspects and scope of fisheries research. A wide range of reports, publications and strategic plans covering socio-economic, biological, environmental and regulatory issues of Lake Victoria fisheries in general and Nile perch in particular were accessed. Where necessary, copies of the literature and reports were made for further processing. Literature review enabled us to substantiate in retrospective the activities and challenges of the channel.

Focus group discussions and individual interviews: Focus group discussions refer to discussions that are limited to a particular topic(s) and discussed by a group of people (Stewart and Shamdasani, 1990). A typical focus group interview involves 6-12 people who discuss a particular topic. Focus group discussions were only possible with fishermen because it was difficult to find middlemen in numbers that could warrant a group discussion. Some middlemen however participated in joint focus group discussions with fishermen. Both focus group discussions and individual interviews were semi-structured (Yin, 2003; Stewart and Shamdasani, 1990). On one hand, unstructured interviews are suitable if the objective of the research is to learn about issues that matter most to the group (Stewart and Shamdasani, 1990). On the other hand, a more structured approach may result in the group discussing only what it may feel important for the researcher. In order to balance the research standards and discover what was important for the groups with regard to improving sustainability and quality, a semi-structured approach was deemed most suitable.

Both group and individual interviews started with a basic introduction of research goal before specific issues about sustainability and quality were discussed. This is consistent with recommendations for case study research (e.g. Stewart and Shamdasani 1990). Topic lists on issues to discuss were prepared in advance of each field visit as recommended by Stewart and Shamdasani, (1990) and Yin (2003). Interviews were transcribed for further processing and issues requiring clarification were verified on repeat visits. Each group discussion lasted for 2-3 hours which was reasonable according to Stewart and Shamdasani (1990) who note that typical group discussions last 1.5 to 2.5 hrs. Individual interviews lasted for 30 - 45 minutes consistent with Yin (2003) who note that interviews may take about 1 hour. In total, we conducted 18 focus group discussions involving 4-10 people per group in 8 beaches, 31 interviews involving 1 to 3 people and 16 key informant discussions with the BMUs. One factory was visited twice and the other two were visited three times.

Personal observations: Personal observation of transaction activities is another way of collecting data on issues that may otherwise not be envisaged from interviews and discussions (Yin, 2003). The observations at the beaches enabled us to identify factors that may be critical for quality and sustainability. Activities and practices deemed critical were explored further through discussions with respective actors. Observations were undertaken each time we visited the beaches either before or after interviews. We observed quality assessment, weighing, packing, and fish sizes. Observations were also conducted at one factory where quality defects as a result of poor fish handling were detected.

3.3.4 Research reliability and validity

A number of steps were taken to ensure reliability and validity of this study. First, the study involved multiple cases, a number of participants and also a number of data sources. Use of multiple cases and sources of information enhances reliability of the results because it enables triangulation of the data (Yin, 2003; Eisenhardt, 1989a). Moreover, substantiating the outcomes of the interviews and personal observations with existing literature enables us to establish a consistent trend of issues and events in retrospective leading to the prevailing state of affairs for the channel.

Furthermore, a case study protocol (see Appendix 3.1) and advance topic lists ensured that all critical issues were covered thereby minimising omission errors (Yin, 2003). In addition, comparing notes with the research assistant enhanced the reliability of the transcribed interviews. During discussions and interviews, participants qualify their responses by certain contingencies or expressions which according to Stewart and Shamdasani (1990), amount to ecological validity not found in constrained survey research. At the end of the case study, preliminary results were presented to stakeholders from KMFRI and Fisheries Department (FD) in Kenya who gave feedback. According to (Yin, 2003), stakeholder feedback is another way of ensuring reliability of case study results.

3.3.5 Data analysis

Data from different landing sites and also from different fishermen and middlemen were examined for differences in their views, perceptions and activities with respect to sustainability and quality. There were no major differences to warrant disaggregate analysis, i.e., separate analyses for the eight beaches. As we will elaborate with the channel structure; isolating one channel from another is difficult due to the intricate nature of the supply networks. Hence, we present an aggregate analysis.

3.4 Results

This section presents the results. First, we analyse the structure of the channel showing the main actors as well as the flow of fish from landing sites to processing factories. Second, a brief description of how transactions from fishermen through to processors are coordinated is given. Third, we present channel relations focussing on power and dependence relations that have implication on the way the terms of transaction are determined. Fourth, we will analyse quality. We present the challenges that fishermen face in trying to improve the quality of their fish. We focus on the factors contributing to quality deterioration and how fishermen cope with them. Fifth, we analyse sustainability, i.e., how the activities of the fishermen, middlemen, processors and the fisheries department may impact on sustainability of the fisheries. Finally, we present the challenges that local communities face in accessing Nile perch fish for food security.

3.4.1 Channel structure

Fish flows from different beaches through intricate networks of middlemen to fish processors (Figure 3.1). Although the exact number of Nile perch fishermen is not known, the recent estimates indicate that there are about 37000 fishermen in the Kenyan part of Lake Victoria (LVFO, 2000). Middlemen operate in different layers. There are middlemen that buy fish from fishermen (referred to in this study as primary middlemen) and those that buy fish from primary middlemen (secondary middlemen). Secondary middlemen sell to those contracted by processors (contracted middlemen). At the time of the study there were only 6 factories operating in Kenya. Moving from fishermen down to processors the channel is an oligopsony, i.e., with many suppliers and very few buyers (Scherer and Ross, 1990).

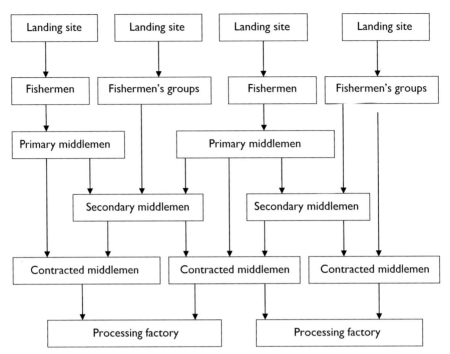

Figure 3.1. The Structure of the upstream of Fresh Nile perch channel in Kenya (Field study 2004/2005).

In some beaches, fishermen work in groups - the so-called associations or cooperative societies which replace the primary or secondary middlemen. The presence of the different layers of middlemen is attributed to the structural changes that followed the expansion of export markets and the decline in fish production. Following the expansion of export markets many actors in the artisanal processing and marketing sectors were displaced.[1] With declining supply and increased competition, contracted middlemen tend to buy fish from a number of beaches. This makes it impossible for them to be present at all beaches when fish is landing. As a result, they either buy fish from (or employ) secondary middlemen. Similarly, some secondary middlemen buy fish from different landing sites. They also buy fish from (or employ) primary middlemen who reside within the vicinity of the beaches and buy fish as it lands. With these supply networks, fish from almost all landing sites go to almost all factories, which makes it difficult to isolate one channel from another for disaggregate analysis. The fact that the same middlemen and hence processing factories buy fish from almost all landing sites corroborates the similarities in the way transactions are undertaken at the landing sites. Despite the intricate supply networks, there are two distinct channels from landing sites, i.e., the domestic channel and the channel to the processing factories. All good quality Nile perch bought by middlemen

---

[1] Abila and Jansen (1997) estimated that for the 2400 jobs that were created in Kenya factories about 15000 were displaced in the traditional fish processing and marketing sectors.

goes to processing factories from where it largely goes to export markets and high class domestic markets. The poor quality fish and juveniles is sold to local consumers.

3.4.2 How transactions are governed

This section outlines how transactions from fishermen through to processing factories are governed.

Fishermen - middlemen transactions: Transactions between fishermen (or fishermen's groups) and middlemen are predominantly based on informal arrangements, i.e., without written agreements. A common feature of such transactions is the provision of informal credit by the middlemen (in cash or kind, e.g., fishing gears) in exchange for exclusive fish supply. Some fishermen and middlemen also help one another on other issues such as in times of sickness, funerals, food shortages or other emergences. Such assistance may however be tied to fish supply especially if fishermen are the beneficiaries. Nonetheless, such assistance compels beneficiaries to reciprocate a helping hand when need arise. Repayment conditions differ across relations (see Box 3.1 for two examples).

Except for the exclusive fish supply to credit providers, other conditions of the credit such as repayment period and total amount to be repaid were not very clear probably signifying the informality of the credit arrangements. Such quasi-credit contracts between fishermen and fish buyers are not unique to Nile perch channel, Platteau and Abraham (1987) report similar arrangement in India. It was however clear for both fishermen and middlemen that credit repayment could end if gears were worn out, stolen or rendered dysfunctional in one way or another. And, depending on the circumstances surrounding the loss or destruction of the gears, it could also end the transaction relationship. At this point, some middlemen reported that sometimes fishermen cheat and declare that gears have been stolen so as to end the loan

---

Box 3.1. Repayment for fishing gears and equipment under informal credit.

1. One middleman had given a boat engine, a boat and fishing nets to 3 fishers. Each time the fishers landed fish, he (the middleman) got 15% of the catch in payment for the engine, 10% for the boat and 5% for the fishing nets. The remaining 70% of the catch was then shared equally among the 3 fishers and the middleman himself. The middlemen was in turn the only buyer of the fishers' share of the catch (Interview at Marenga landing site)

2. Another middleman had 8 fishers to whom he had given 3 boat engines, 3 boats and nets. In return, the middleman got one day's catch for the engine; another day's catch for the boat; and the other day's catch for the nets. The fishers then got the other three days catch which they shared among themselves. They rest one day a week when they repair the fishing gears or go to church. Even in this case fishers sold their share of catch exclusively to the middleman (Interview at Obenge landing site)

---

repayment. While none of the fishermen admitted to cheating in that respect, they admitted that they have lost fishing gears sometimes at gun point in the lake.[2]

Middlemen - middlemen transactions: Transactions among middlemen were also informal. Primary and/or secondary middlemen sometimes get cash advances to help them buy fish. They also obtain cash to meet other social needs. Repayment is predominantly in form of fish supply. To that extent primary or secondary middlemen tend to be tied to one another and/or to contracted middlemen.

Middlemen - factory transactions: Transactions with factories are mainly based on written contracts although some middlemen reported to have oral contract. Generally, contracts outline identification particulars for middlemen, the beaches from where the middlemen would be buying fish, minimum fish sizes, fish price, ice to fish ratio, fish quality and use of factory resources. Some minor differences exist in the conditions for the provision and use of factory resources such as trucks, ice, fuel and sometimes cash advances. For example, one factory was demanding advance but refundable payment for security of the trucks and ice. The advance payment was refundable if middlemen did not misuse ice or fuel or damage the trucks. Another factory was charging for ice or fuel deemed misused.[3] Middlemen were expected to use a certain amount of fuel depending on the landing sites where they buy fish. They were also supposed to buy fish according to the amount of ice in the truck. So fuel and ice used in excess of expectation was considered misused. Another factory was contributing a proportion of truck driver's allowances and fuel costs. These conditions also varied among middlemen from the same factory. For example, one factory was asking for advance payment from new or old but untrustworthy middlemen.

3.4.3 Channel power and dependence relations
On the basis of our interviews and observations, this section discusses the sources of market power and how they affect the way transactions at the landing sites are undertaken, (1) access to technological tools for quality improvement, (2) access to price information and (3) interlocked fish/credit markets.

Access to technological tools for quality improvement: With declining fish production, one would expect that fishermen who have fish - the backbone of the transactions - should be able to competitively bargain for better terms of transactions. This is not the case with fresh products in absence of the storage facilities coupled with the perishability of the product. In the Nile perch channels, contracted middlemen always have access to refrigerated trucks for fish transport and storage while at the landing sites. These middlemen then take advantage of lack of tools for quality improvement at landing sites to impose low prices. This works to

---

[2] Incidences of gear theft with emergence of armed robbery are said to be increasing in the lake (Abila, 2000; Geheb and Crean, ed., 2000; Geheb, 1997).

[3] At the time of the study the factory charged KSh4/kg for ice deemed misused and KSh40/litre of fuel deemed used in excess of the allocated mileage.

their advantage because fishermen are left with two choices to either accept the low prices or risk more fish spoilage if they do not sell fast. We illustrate this in next section with price information.

Price information: Daily fish prices are often not known to fishermen until when middlemen arrive at beaches. Upon arrival at the beach, middlemen inform the BMU about the price. The previous price could be sustained, or lowered or increased. There is some negotiation between the BMU or fishermen's group and middlemen. The set price remains the same for the day for the fish that is deemed good quality for the export channel. Although some middlemen reported that processors also sometimes change price without notice, they often have price information from their contracts before they buy fish at the landing sites. When consulted, two of the processors noted that they always notify middlemen of pending price changes.

Fishermen also reported that middlemen take advantage of information asymmetry and lack of storage facilities, to collude and keep fish price at the beaches low. One middleman interviewed at one landing site acknowledged that middlemen sometimes collude to keep prices low. This is consistent also with Owino (1999) who noted that "*agents* (middlemen) *unite in order to keep the prices low.*" During one of our visit to one of the beaches, we encountered an incident where middlemen forced fishermen to lower the prices (see Box 3.2).

Although prices at the landing sites tend to be fixed for each day, they vary widely across beaches and time.[4] Whereas variation in time may reflect seasonality of fish production and the continued decline in fish production, variation across beaches was sometimes not consistent with economic reasoning. For example, prices were sometimes higher in beaches far away from the factories where by implication transport costs could be high. Prices were sometimes also higher in beaches patronised by fewer middlemen where competition was low. One would expect prices to be higher where landing was high in order to compensate for shorter waiting times. But this was not always the case as sometimes prices were almost the same in beaches with high and low landings. As all the fish deemed good for export channel was sold at the same price, there were no price variations as a result of fish quality. Similarly, quality variations may not explain the price variations across beaches because all beaches sell fresh fish and the mode of quality assessment are entirely the same.

Interlocked credit and fish markets: Fishermen's vulnerability to price fluctuation can also be attributed to the interlocking of fish and credit markets. For example, even if fishermen know that prices at other beaches are higher, those that have credit ties with middlemen cannot sell fish at other beaches because of the exclusivity of fish supply to middlemen. Fishermen tied to middlemen through informal credit especially fishing gears and boats also have low bargaining advantage. For example, bargaining for better terms may lead middlemen to withdraw their

---

[4] Between February to December 2004, fish prices in the case study beaches almost doubled, i.e., KSh 53 - 96/kg. In Feb, (2004) prices across beaches ranged from KSh53-67/kg and in December (2004), prices across the same beaches ranged from KSh 70- 96/kg).

Box 3.2. Agents walk-away to impose low prices.

Upon arrival at one of the beaches, we found 3 contracted middlemen from three factories from Kisumu, Nairobi and Migori and the BMU in a meeting. After a while the meeting disbanded without an agreement which gave us time to talk to both the middlemen and the BMU secretary.

The bond of contention was fish prices. When middlemen arrived at the beach they told the BMU that the factory prices had gone down such that they could only buy fish at KSh55/kg down from KSh67/kg on the previous day. In protest fishers refused to sell fish. After some time of negotiation fishers offered to sell fish at KSh65/kg but the middlemen insisted that they could only afford KSh55/kg. The meeting then disbanded.

During our interview with the middlemen, they claimed that prices at their respective factories had been reduced overnight such that they could not afford to pay what fishers wanted. When asked to explain the coincidence that the three factories lowered the prices to the same level overnight, "It usually happens, it's not new, after all these factories belong to the same Indians who collude to fix the prices," was the explanation from one of the middlemen while the other two nodded their heads probably in concurrence. After talking to us the middlemen left the beach without buying fish.

In a separate interview, the BMU secretary reported that they could have believed that prices at the factories had reduced if the middlemen had shown written evidence from the factories. As the day went by no other middlemen came to the beach as we hanged around to establish how the walk-away could end. Meanwhile fishers were speculating that other middlemen would not come because "they must have phoned each other." We left later in the afternoon before other middlemen came.

On our follow up visit to the beach we were told that the same middlemen came back later in the afternoon only to buy fish at KSh57/kg while some fish had been spoilt. We could not estimate how much fish was spoilt.

fishing gears and look for other fishermen. As one middleman said, "*If a fisher is not happy with the conditions for using my gears, I can just withdraw them and give to any other fishermen. There are many fishermen who want fishing gears and boats*".

This probably explains why ownership of gears given under the so-called informal credit is often not transferred and repayment period was never clear. Similarly, middlemen who use factory resources may not easily take advantage of price differences between factories also because of the condition of excusive fish supply.

The problems with interlocked fish/credit markets on fishermen have previously been noted. For example, Bokea and Ikiara, (2000, p 17) noted: *"local fishers have gradually lost control over the means of production as well as processing, pricing and marketing to industrial investors. In pricing, local fishers have no say because of lack of storage facilities, the high perishability of fish and the pressure from the credit relationships."* Similarly, Henson and Mitullah, (2003, p 72) note: *"Given that there are virtually no chilling facilities at landing beaches, processors have to provide ice and insulated trucks and ... whilst they incur significant costs as a result, it also enhances their market power over fishermen".*

In spite of these seemingly exploitative informal credit relations and in absence of alternative sources of credit (e.g., formal credit institutions for small scale fishermen), fishermen especially those that cannot afford to buy their own boats and gears acknowledge that such relationships keep them going. This is consistent with Platteau and Abraham (1987) who note that although such credits tie beneficiaries to creditors, they both benefit from them either as an insurance against any eventualities or securing access to output respectively.

In short, lack of access to quality improvement facilities and price information, and interlocked fish/credit markets are critical factors influencing the way upstream transactions are undertaken. The problems are exacerbated by the structure of the channel that enhances the buyer concentration and hence, dominance over suppliers. Such problems can be dealt with, for example, if there are formal micro-credit institutions that would enable fishermen to obtain loans for fishing gears. Such institutions would not only disentangle fishermen from interlocked markets but also enhance their bargaining power. The information asymmetries could be solved if there was a governing body such as cooperative society and fish marketing board that could oversee the fish marketing activities including dissemination of market (price) information, negotiating prices or searching for better fish markets on behalf of fishermen. Middlemen could have a similar organization to negotiate with processors.

3.4.4 Quality
On the basis of our interviews and observations this section presents quality assessment at landing sites and then at the factories.

Fish quality at landing sites: The first evaluation of the raw Nile perch is done at the moment of landing at the beaches using sensory assessment by checking the colour of gills, eyes, skin and the firmness of the fillet. After assessing, fish is declared good or bad quality. The good quality fish is sold to middlemen for export channel and the bad quality fish often fails the sensory tests for quality and is sold to domestic traders who sell it in domestic markets. However, the sensory techniques do not always give consistent results. For example, fish could have gills or eyes with a colour indicative of poor quality while the fillet could still be firm or vice versa. Also fish could have colour of the skin and eye on one side depicting good quality while on the other side depicting poor quality. Inconsistent quality indicators sometimes become a source of conflict when middlemen reject fish that fishermen feel it is of good quality. Fishermen reported such problems occur when there is more supply and middlemen raise quality standards. Deterioration of fish quality is aggravated by a number of factors such as, (1) lack of tools for quality improvement, (2), period from fish catch to landing, (3) period between buying and delivery to the processing factory and (4) physical handling of fish.

- *Tools for quality improvement*: There were no cooling facilities in all the beaches studied.[5] In absence of these facilities fishermen and middlemen improvise means to minimise fish spoilage. For example, they keep fish in shades with mouths and gills open for air circulation. During the interviews, it was claimed that such techniques extend shelf life for about 6-8hours. Also middlemen remove gas from the berry of the fish which they also claim delays the onset of spoilage. It was beyond the scope of this study to verify these claims.

- *Period from fish catch to landing*: Fish take long before being put to ice because fishermen take long to land. Fishermen rely on manual or wind propelled canoes and so time to land is sometimes dependent of the strength and direction of the wind on the lake. Fishermen also take relatively longer to catch enough fish due to declining catch. In the process, the fish first caught is subjected to longer periods without ice under high tropical temperatures. On average fishermen estimated to take about 5–8 hours from fishing to landing. Fishermen estimated that about 10-30% of fish could be spoiled by the time they land depending on the season with the high spoilage in rainy season. Quality deterioration was also attributed to fishing methods. For example, fish caught with hooks was reported to have fewer or no scratches than fish caught with nets. The fish caught with nets tend to have bruises that are inflicted when fish gets entangled in the net. Although we did not observe the fishing itself, the damages on skin and fillet between fish caught with hooks and nets were evident.

- *Period from buying to delivery at factory*: On average, middlemen take about 2-5days to buy and deliver fish to the factories depending on the season. This was due to declining supply that intensifies competition at landing sites. Logistically, middlemen do not institute the principle of "first come first serve" when buying fish at landing sites. Whether or not they supply fish to the same factory, middlemen have independent contracts such that they compete to source fish. Consequently, a number of trucks load fish at the same time – each loading less fish per day even if total landing at the beach could be enough for one or two trucks per day. As trucks stay long ice melts - thereby reducing fish to ice ratio. Fish spoilage could be inevitable even if fish is in ice.

- *Physical handling*: Sometimes fishermen kill the fish by smashing the heads or body or let it die in the open air. Rough handling of the fish results in bruises. It was also common practice for those weighing or loading fish in the trucks to throw fish. Those seen throwing fish argued that throwing fish in slanting position or on ice does not damage the fillet. While these claims could not be verified, throwing fish could reflect limited understanding of how such handling could affect say texture of the fillet. Unfortunately, the damage resulting from mishandling fish may not be visible at the landing site though after several

---

[5] According to Henson and Mitullah, 2003 out of the 300 landing sites in Kenya only 30% have weighing shades (locally known as Banda), 9% have electricity, 5% have a fish store, 0.3% have a cold room, 20% have access to all-weather roads.

hours or even days the detrimental effect emerges (e.g., at the factory). Some middlemen acknowledged that they get as much as 10-20% rejects at the factory even when they undertake sensory quality assessments for all the fish before they buy at the landing sites.

In summary, fishermen face constraints in terms of access to tools for quality improvement. For example, ice is virtually not available. Even if it would be available, not many fishermen would be able to take it to the lake due to the limited sizes of the canoes. To minimise delays due to wind, fishermen may use boat engines. However not all fishermen can afford to buy a boat engine. Furthermore, it is not always that delay to land is due to wind problems. Some delays are due to lack of adequate fish to warrant landing. Hence, there may not be one solution to the quality deterioration problem at landing sites. In this respect, catching method, handling and storage regime of the Nile perch are important quality control points.

Fish quality at the processing factories: All the factories follow HACCP as a matter of mandatory requirement by the export markets. The factories also assess fish quality by similar sensory methods as those used at the landing sites. Defects can be related to the condition of the fish flesh, appearance, which include colour defects (bruises, bloodspots) and dehydration (frozen storage defects). The Nile perch processing industry uses 3 quality grades (A, B and C) of the whole fish and defines these grades with clear descriptions of the general appearance, eyes, gills, odour, skin, smell and texture (Table 3.1).

Only grades 'A' and 'B' are accepted and C is rejected. With this quality assessment, middlemen tend to be paid after verification of quality sometimes after filleting. Factories estimated that

Table 3.1. Fish grades in the processing factory for Nile perch.

Sensory	Grade A	Grade B	Grade C
General appearance	Bright with metallic luster	Some loss of metallic luster	Bloom completely gone or bleached
Eyes	Bright translucent	Dull, slightly sunken	Dull and sunken
Gills	Bright red to slightly pinkish red	Pinkish red to brownish red	Brownish red to brown or grey
Odour	Fresh odour Faint sour odour	Faint sour or fishy odour to medium sour odour	Medium to strong sour odour
Skin	Natural brilliance metallic shine	Greyish appearance loss of metallic shine	Yellowish slime greenish fins
Texture	Firm and elastic	Moderately soft	Generally soft, flabby and pits

Source: Fish processing factory in Kenya

about 20-40% of the fish rejected at the factory is often due to poor handling such as throwing or stepping on fish that cause crumbling and/or discolouration of the fillet.

In short, processing factories assess fish quality from outside and also inside after filleting. The quality damages arising from poor handling become visible at the factory. Unfortunately, middlemen may not easily trace the source of poor quality fish back to fishermen because they do not label the fish per source and also the fact that they sometimes take long time before delivering fish to the factory, they may not justify that the spoilage is due to poor handling by fishermen.

3.4.5 Sustainability of the fisheries

In Kenya, sustainable utilization of the fisheries resources is regulated by law - the Fisheries Act (Kenya Government, 1991). The Act outlines the type and size of gears for specific fish species in specific water bodies. It also designates fishing places to protect breeding zones and the minimum fish size that should be landed, bought or sold. Furthermore, the Act outlines conditions under which one would be eligible to fish, trade or process fish products. Failure to enforce the Act is an offence punishable by law. We examine how fishermen, middlemen and processing factorises, and regulatory institutions enforce sustainable practices as stipulated in the Act and the challenges they face.

Fishermen: According to the Fisheries Act of Kenya fishermen are expected to catch Nile perch using gill nets of at least 12.7cm when stretched diagonally or hook number 5. Using these types of gears is expected to catch recommended fish sizes of 50-85cm. Nile perch of less than 50cm (i.e., 1.5kgs) is considered a juvenile and therefore should not be caught. If a fisherman lands Nile perch of less than 1,5kg, it means that he uses small gears.

The fisheries degradation is, among other factors, attributed to increasing use of destructive fishing gears that catch undersized fish. The use of destructive fishing gears and methods is further attributed to: (1) declining fish catch; (2) limited alternative sources of livelihood in other sectors of the economy; (3) high prices for recommended fishing gears; (4) lack of justice when use of bad gears is detected and biased enforcement of the law against fishermen. As fish stocks decline, fishermen argue that they are compelled to use bad gears otherwise they would not catch any fish and in a constrained economy, they do not easily find alternative sources of livelihoods. Some fishermen also argue that recommended fishing gears are relatively expensive (estimated at US$30 per fishing net) compared to bad gears that were reported to be relatively cheaper (estimated at US$10 per fishing net) and also readily available. In addition, some fishermen argue that when non-compliance with sustainable practices is detected, justice is not often done and if it is done it is always biased against fishermen and not middlemen or processors as one participant noted: "*You will never hear that a truck has been impounded or a factory has been closed or fined for buying or processing undersize fish – they know them*".

Middlemen: Middlemen are also restricted by the Fisheries Act to buy only the fish sizes legally recommended. But middlemen transfer the blame for disobeying the Act arguing that they buy what their customers (i.e. processors) demand. Whereas this might make sense from a market – orientation perspective, they likely take advantage of the weaknesses in the enforcement mechanisms that they may not be punished for the undersize fish they trade in. The middlemen also blamed some fishermen who they claimed, use small fish as a bait to sell the big ones. This claim was confirmed during one discussion when one member of the BMU said *"if they* (middlemen) *only want big fish then who will buy the small ones?* Although the middlemen's contract with factories explicitly indicate the minimum fish sizes (mostly 1kgs), observations and interviews revealed that violation of the size was common, systematic and more rewarding in terms of price margins. For example, an interview with one middleman illustrates how rewarding trading in juveniles could be (See Box 3.3).

In short, middlemen like processing factories, are profit-oriented and it could be in their interest to trade in small fish to enhance their economic gains. Hence, transferring the blame for buying and supplying undersized fish to fishermen, processors or ineffective enforcement by the FD could just be a cover-up to maximise their short-term economic gains. There is need therefore to create an environment where middlemen feel responsible to protect the fisheries and fishermen are not viewed as the only ones responsible to protect the fisheries because they too have private interests to fulfil. Fishermen need to be supported to protect the fisheries and one way to do that is for other parties like middlemen and processors to be held accountable for irresponsible behaviour and for responsible fishermen to be adequately rewarded for their contribution to sustainability.

The Fisheries department: The Fisheries department (FD) is legally mandated to enforce sustainable management of the fisheries. To do this, the FD issues licenses to fishermen, middlemen and processors that authorises them to undertake their respective activities according to the law. The FD is also supposed to take action when actors violate the terms of

---

Box 3.3. Price margins from undersize fish (Interview with a contracted middleman).

One middleman was buying undersize fish from 0.5 – 1kg at KSh40/kg. He was also buying fish of more than 1kg at KSh65/kg. During our interview with him the same day, he reported that at the factory fish will be sorted into three groups i.e. those less than 1kg, 1-5kgs and over 5kgs. Then, all the fish of less than 1kg will be sold to the factory at KSh55/kg resulting into a price margin of KSh15 (55-40)/kg. Fish of 1-5kgs will be sold at KSh75/kg creating a price margin of Ksh10 (75-65)/kg. Finally fish above 5kg will be sold at KSh77/kg creating a price margin of KSh12 (77-65)/kg. This shows that buying and selling undersize fish give higher price margins.

Note KSh. = Kenya Shillings (local currency in Kenya)

---

their licences. During the discussion we examined fishermen's and middlemen's perception about how the FD undertakes their activities. We also sought the views of the FD staff.

There were mixed perceptions about the effectiveness of the FD in protecting the fishery. For example, participants interviewed wondered why the FD does not close down companies that manufacture illegal gears. One fisherman however, differed with this view that companies make illegal gears. The fisherman argued that "*It is not always true that companies make illegal gears, but that we* (fishermen) *use the good gears in wrong places or on the wrong fish types*". This argument was in line with the fact that the Fisheries Act specifies different types of gears for different fish types in different water bodies. Hence a gear that is good for one fish type in one water body may not necessarily be a good gear when used in another water body. However, this was an isolated remark. Common remarks among fishermen and middlemen suggest a number of negative perceptions about how the FD undertakes their responsibility. Such remarks suggest anything from leniency in dealing with cases of non-compliance to bribery.[6] For example, one participant at a group discussion wondered that: "*They* (FD field officers) *know who has illegal gears, why not just get and destroy them?*" Another participant echoed noting that: "*The fisheries officers can arrest a fisherman for having an illegal fishing gear today but few days or weeks later, you see him back in the water using the same gear, what does that mean?*"

One way to control and monitor access to and proper use of the fishery, the Fisheries Act mandates the FD to license all fishing gears. The extent to which gear licensing has been used for the intended purpose has been questioned by both fishermen and middlemen. The FD is blamed for emphasizing more on the monetary benefits than sustainability and livelihoods of the local people who depend on the fishery as the following remarks suggest: "*When government officers come to licence fishing gears, they licence every fisherman they find, whether they have good or bad gear, they do not care as long as one has the money to pay for the licence fee,*" recounted one fisherman. These remarks are in line with Abila (1998) who wondered "whether gear licensing is just another way of raising government revenue or really to regulate access to the fishery." We sought the views of FD field assistants over this allegation and one field assistant said: "*I cannot manage to physically check thousands of hooks, nets one by one. I am responsible for many beaches and it is not practical.*" This remark essentially corroborates the allegation that even bad gears tend to be licensed for whatever reason. Geheb (1997) noted that the FD has few field assistants who are responsible to monitor the activities of many fishermen.

In short, the FD involvement in the fisheries management has generated a lot of arguments. There are reports that before the FD involvement in the fisheries sector, i.e., before Nile perch era, the fisheries were not only sustainably managed under local leadership but the benefits from the fisheries were equitably distributed (Henson and Mitullah, 2003). The fisheries were

---

[6] Previous cases of bribery have been reported (Abila, 2000; Owino, 1999; Geheb, 1997). For example, Owino (1999) found that fisheries scouts obtained about $250 each per month i.e. about 5 times their monthly wages (US$50-60) in bribes after allowing fishermen to use illegal gears in breeding and spawning areas.

managed by local institutions which enforced who were entitled to fish, in which seasons, which fishing zones, types of fish gear that could be used, size of fish that could be landed. The FD intervention disrupted the role of the local institutions in fisheries management following the expansion of export markets which has been viewed as being more commercial- than sustainability- oriented. Geheb (1997) note that the unregulated 'free for all fishing' under the government watch to maximise economic gains from export earnings has contributed negatively to sustainability. It might therefore be necessary to revive the role of the local norms and institutions that "sustainably" managed the fisheries or at least to develop alternative mechanisms that should bring the local institutions and actors back into responsibility for managing the fishery in collaboration with other stakeholders. Fortunately, some local institutions – the BMUs are already taking up responsibility in Kenya.

3.4.5.1 Emerging opportunities - the BMUs engage in sustainable fishing practices

Despite the poor perceptions against the FD, some fishermen in few beaches have resolved through the BMU to implement sustainable fishing practices. For a few years now, fishermen in such beaches have been using recommended fishing gear. In order to help the fishermen obtain the good gears, the BMUs run a self-help fund to which fishermen contribute a small proportion of their daily fish sales. Any fisherman facing problems such as loss or wearing out of gear or other emergency needs, obtain a loan from the fund. This organization is particularly successful at one of the 8 beaches. At two other beaches, efforts were underway to organise fishermen. In an interview with a secretary of one of the few BMUs, we asked how it worked out that fishermen at the beach comply with the beach resolution to use good gears. In response the secretary had this to say:-

"We are all fishermen (including himself), the lake is all we have to live on. We can make things change if we want to. Hiding behind poverty to destroy the lake will not help us. We (the BMU) made fishermen to understand and together we resolved to protect our resources."

A fisherman at another beach where efforts are also underway to promote sustainable fishing had similar comments: *"Fisheries officers are employed, they are assured of their salaries whether they help us deplete the lake or not. If there is no fish in this lake, they will be transferred to another lake. The same with the Indians* (factory owners are mostly *Indians) they can relocate and continue their businesses in other lakes or countries. No one will transfer our problems to anywhere and the lake is what we have to survive on."*

These are signs of emerging responsibility to enforce sustainable utilization. No matter how small in effect compared to number of the beaches and the size of the lake, it might be the starting point to support sustainable fishing. Notwithstanding their newfound sense of responsibility for sustainable fishing, the fishermen in these few beaches, like their colleagues in other beaches, lack facilities such as ice to keep fish fresh. They therefore suffer equally from the dominance of fish buyers, for example, in terms of pricing. Consequently, their contribution to sustainability is, in our view, not adequately rewarded.

Therefore, to sustain this newfound sense of responsibility and motivate more fishermen to engage in sustainable practices, there is need to support and reward responsible fishermen. While punishing irresponsible behaviour is another way of motivating responsible behaviour, it may not always be effective, after-all, detecting and punishing irresponsible behaviour might not always be easy without effective monitoring. As Geheb (1997) note, fishermen face the possibility and vulnerability of being caught, but what matters is that "benefits derived from the use of illegal gear outweigh the costs after being caught." Hence rewarding responsible behaviour would be a better way of dealing with irresponsible behaviour. The question, of course, is what would motivate them.

3.4.6 The local communities access to Nile perch for food security
The declining fish production and increasing competition between domestic and export channel for Nile perch raises challenges for the rural communities to secure it for their food security. The Nile perch started on a promising note. Following its successful introduction, Nile perch production led to unprecedented socio-economic benefits in terms of food security and income for the rural communities and fishermen in the region. For example the rural communities in Tanzania crowned Nile perch as "*Saviour*" (Gibbon, 1997) and in Kenya, the Luo –the main fishing tribe after appreciating the generous Nile perch food rations said that "*Mbuta* (Nile perch) *taught women how to feed their Husbands well,*" (Geheb, 1997; p66).

At the moment of declining Nile perch and fisheries in general, the implications are far reaching. Nile perch fish is one of most scarce product in the domestic markets in Kenya. Over 90% of good quality Nile perch goes to export markets and the remaining goes to the "high class" segment of the domestic such as hotels where ordinary people cannot afford. Previously when Nile perch was new and plentiful, failure to eat it was a matter of choice among other plentiful species. At the moment however, scarcity is largely due to low production and high competition between domestic and export markets. The latter having an advantage over the domestic markets, domestic traders that once proliferated as a result of Nile perch boom are now reduced to processing only what is rejected on quality grounds, the juveniles and the by-products from the processing factories such as skins, skeletons, trimmings, fats, among others. Yet, supply of these quality-rejects and by-products is not without hurdles. For example, it costs domestic traders - predominantly women waiting the whole day at landing sites yet without certainty that they would get any rejected fish. Although it is almost always certain that some Nile perch would be rejected at the landing sites, it is not always enough for each woman to have a piece. Depending on the weather, as one woman indicated that, "*Our business becomes slightly better during rainy season when there are more rejects*". Abila (2000) notes that only when factory trucks fail to reach the landing sites either due to bad roads or external problems like export ban (like once imposed by EU) could women freely buy Nile perch.

Whereas failure to secure rejected Nile perch at the landing sites is due to inadequate fish rejects, women who trade in factory by-products face competition from the animal feed manufacturing companies. As a result, women also wait for long at the factories sometimes not being sure if they would buy the by-products. It is estimated that about 59% of the frames go for animal feed manufacturing per day further depriving the domestic market (Henson and Mitullah, 2003). Ironically, in early 1980s, Nile perch skeletons were considered as a waste and factories incurred costs to dispose of them (Abila, 2000). It was not until mid 1990s when people started eating the skeletons as a result of the shortage of Nile perch and the increased prices of fish. The skeletons were then considered as a "poor man's food" implying that only those who could not afford the good quality Nile perch could eat the skeletons. Abila (2000) estimates that before advanced filleting technologies were introduced about 10-20% flesh could remain on the skeletons that people could eat. But with improved filleting technologies that remove almost of the flesh, the skeletons become "*too naked*" to be of any further food value for human consumption (personal field observations). But even then local traders scramble for them and by late 1990s, the value and prices of the skeletons had increased beyond what other poor people could afford (Jansen, 1997, see also the *Darwin Nightmare* documentary -Sauper, 2004).[7]

In short, at the moment the welfare of the local communities around the Lake Victoria regions is another challenge that the Nile perch channel may have to face upfront. It is much likely to be even worse if the decline in fish stocks is not checked. There is a danger that the "social decay" that the rural communities have been subjected to as a result of the commercialization of the fishery and declining fish stocks may likely continue if nothing is done against it.

3.5 General discussion and implications

The objective of this chapter was to analyse the situation in which fishermen and middlemen operate in order to understand the market failures that they face, how they cope up with them and how they could be addressed. On the overall, the chapter has unravelled a number of technological, socio-economic and institutional constraints and also emerging opportunities to promote sustainable and quality- enhancing practices.

The results show that fishermen's position in the channel is compromised by the structure of the channel that gives structure power to buyers, lack of price information and interlocked fish/credit markets. While it may be difficult to change the structure of the channel, power of the fishermen can be changed through minimising information asymmetries especially over price. There is need therefore to create market information systems and institutions

---

[7] *Darwin's Nightmare* is a documentary surrounding Nile perch exports to the EU. It captures the complex issues of poverty, food insecurity, armed conflicts and HIV/AIDs, among others, around Lake Victoria. The documentary, however, needs to be interpreted with caution, as the IUCN and the Lake Victoria Fisheries Organization in an open letter to the author posted on IUCN website argue that the documentary distorts the reality of Lake Victoria fisheries (www.iucn.org/IUCN News – December, 2005).

through which price information could be communicated to the fishermen rather than the fish buyers. For example, use of modern technologies such as short messages through mobile phones with factory prices to subscribed BMU's or fishermen. Such institutions could be through the establishment of a fish marketing institution that could coordinate information flow to fishermen. The use of mobile phones by fishermen to access market information is being used in Asia (see, The Economist, 2001a&b, 2005a&b). However, this may only work better if fishermen were free to decide where to sell fish. In the Nile perch channel where some fishermen are tied to fish buyers that information may have limited impact. This means that first fishermen need to be disentangled from interlocked fish/credit markets. That can be done by making fishing gears affordable or establishing micro-credit institutions with reasonable conditions where fishermen can obtain loans for fishing gears.

This chapter also shows a number of factors that may affect quality such as lack of proper knowledge on fish quality attributes reflected by poor handling; lack of cooling and storage facilities essential for fresh and perishable products, the type of fishing gears, and the time it takes before the fish is processed. Although these factors are not necessarily hazardous and therefore, are not taken up in HACCP which is about hazardous aspects (mainly microbial), they are nonetheless important for quality assurance at primary level. These results imply that investment in the tools for quality improvement such as ice or cold rooms at landing sites or investments in larger boats that can carry ice and fishing crew are needed. It may also require even more structural investments such as electricity that is currently not available in all the beaches studied. Poor handling practices could be minimised through educating the fishermen and middlemen on the effect of poor fish handling on quality. More importantly, it may require motivation for example, better prices for better quality. Fishermen may handle fish properly if they know that they would be rewarded for it.

The results show that the use of bad fishing gears is attributed to high prices for the good gears and ineffective enforcement by relevant authorities. This implies that public institutions should improve their effectiveness in enforcing sustainable fishing practices. Further recommended fishing gears should be made affordable to all fishermen. However, neither the effectiveness of public institutions may improve overnight nor making gears affordable to all fishermen might, by itself, guarantee that all fishermen would use good gears. Sustainable practices could be enhanced by a combination of both enforcement and affordability of the gears. For example, by only targeting and punishing fishermen, the FD is implicitly rewarding middlemen and processing factories for buying and processing undersize fish. This in turn may frustrate fishermen who might want to implement sustainable practices because they too have economic motives that may drive them to use bad gears. Above all, while punishment for wrong doing may deter further wrong doing, focussing on the destructive fishermen makes the FD fail to recognise and adequately reward the good work that some fishermen are doing. Therefore, a change in the approach on how sustainable practices are enforced might be needed. For example, fishermen that implement sustainable practices should be rewarded through better prices or through better access to recommended fishing gears, access to storage facilities; access to more profitable markets, i.e., processing factories that may pay them better prices. In that way we believe that other fishermen may follow the good behaviour.

Implementing such incentives should start with fishermen that, through their BMUs, are already implementing sustainable practices. Since the BMUs that implement sustainable practices do so by reasoning with - and motivating - fishermen, the FD may wish to work with the BMUs to motivate more fishermen to implement sustainable practices. Supporting the BMUs may take different forms such establishing micro-credit schemes to enable fishermen in registered BMUs to buy fishing gears at reasonable prices or low interest loans. Such an approach would mean that fishermen who use bad gears under the pretext of high prices or lack of credit facilities would no longer have an excuse. Ultimately, social control by fellow fishermen and the BMUs would be more effective and appropriate.

The declining welfare of the rural communities could be seen as one testimony that marketing mechanisms may lead to overexploitation of the natural resources rendering the poor who depend on them more vulnerable. In this particular case, the export-orientation nature of the Nile perch channel is one of the major factors that continue to deprive the local communities of the fish resources they depend on for food security. Although export orientation could be argued to be an engine for economic growth from the perspective of globalization and market integration, this study shows that export orientation without deliberate intervention to protect the poor may not benefit them. Therefore, there is need to put sound policies, for example, to divert not only the lowest C quality but also the middle (B) quality to domestic markets leaving the highest A quality for the export markets (the question here would be: Do the poor have sufficient purchasing power to buy C or B quality ?). This suggestion is consistent with the arguments by analysts of globalization and poverty who vehemently argue that global market integration *per se* cannot bring economic benefits to the poor without sound country specific pro-poor policies and social protection programs to protect the vulnerable segment of the population (e.g., Basu, 2006; Ravallion, 2006). For instance, Bardhan (2006) suggests that international agencies that preach the benefits of globalization and market integration should have an obligation to contribute to such programs with financial, organizational and technical assistance. So too, local governments and in-country private sector are responsible to protect the vulnerable segment of their society.

3.6 Conclusion and future research

In conclusion, the need for developing and implementing mechanisms that enhance sustainability and quality in the Nile perch channel cannot be overemphasized. Although fishermen, middlemen as well as processors operate at different levels and scale, and may have different interests, one thing is common - they all face risks associated with declining fish stocks and also potential benefits if they adopted sustainable practices. For example, fishermen take long to fish few fish, middlemen take long to deliver fish to the factories and factories fail to utilize their installed processing capacity. Coping with lack of technological tools for quality improvement is for fishermen very difficult, yet the solution i.e., investment in ice or cold storage facilities is beyond what they can afford. Although common property resources are known to be difficult to manage due to the diversity of users interests and lack of private property rights, some fishermen implement sustainable practices which means that the common property resources can be managed by the same commoners. What may be needed is mechanisms and support for fishermen to overcome the constraints they face.

This study is a result of intensive discussions, interviews and observations in eight beaches in Kenya. Although the study gives us a picture about the conditions in which the fishermen and middlemen operate, future research should undertake similar case studies in other beaches especially those in the islands. Having a broader picture of the other beaches would be necessary to develop appropriate invention strategies for the part of Lake Victoria in Kenya as a whole

Appendix 3.1. Summary of the case study protocol

a. Field arrangements
- Review relevant literature from the internet and relevant libraries in Kenya
- Identify and assemble crucial documents and literature for further review
- Visit relevant offices for any other information on Lake Victoria fisheries in general and Nile perch in particular.
- Undertake a familiarization visit to some beaches along the lake

b. Identification of research participants
- Identify key stakeholders from relevant institutions and existing literature
- Establish from relevant institutions and officers about accessibility to the stakeholders

c. Research assistants
- Identify research assistants from the host institution (KMFRI)
- Familiarise the research assistants to the purpose of the research and field procedures

Summary of the issues covered during the interviews and literature review

Important aspect to be established	Source of information
Background to Nile perch fishery and channel	Literature review
Structure of the channel	Interviews
Channel coordination	Literature review and personal interviews
Channel activities and processes	Group discussions, individual interviews and literature review, observation of the transaction process.
Major challenges facing the fishery and the channel	Literature review, interviews with public officers and other channel members
Trends and perceptions of Lake Victoria fishery in general and Nile perch in particular	Literature review, focus group discussions and personal interviews
Knowledge and perception about fish quality and safety	Focus group discussions
Critical factors to fish quality	Personal observations of fish handling practices
The role of various stakeholders in fishery management and fish quality management	Literature review, focus group discussions and personal interviews
Nile perch in the domestic channel	Literature review, interviews and observations in local markets

Chapter 4

Factors influencing preference for sustainability and quality-enhancing contracts among fishermen

4.1 Introduction

This chapter investigates the factors that would influence fishermen's preference for contracts that oblige them to implement sustainable and quality – enhancing practices. In order to do that, the chapter translates the theoretical framework from Chapter 2 and applies it to the field setting introduced in Chapter 3. It has been noted that the market failures that developing nations face motivate small-scale primary producers to engage in contracts to solve them. For example, small-scale farmers engage in contracts to access profitable markets, minimise price risks and uncertainties, minimise information asymmetries, access production inputs, technologies and services, and also access credit facilities, among others (Masakure and Henson, 2005; Grosh, 1994; Key and Runsten, 1999). Using conjoint analysis, this chapter investigates whether addressing these problems would stimulate fishermen to engage in contracts that oblige them to implement sustainable and quality-enhancing practices. The rest of the chapter is organised as follows; the next section reviews relevant concepts leading to the formulation of our hypotheses. The research methodology used to empirically test the hypotheses follows. A discussion of the results and their implications for implementing sustainable and quality-enhancing practices conclude the chapter.

4.2 Concepts

In order to select relevant theoretical concepts, we integrate transaction costs economics and social – and network theory perspectives. Literature on channel governance demonstrate that there is an underlying governance continuum stretching from spot markets at one end to hierarchies at the other, interspaced with hybrid mechanisms such as relational norms and contracts (Williamson, 1985). According to TCE, circumstances that necessitate contracting include specialized investments, information asymmetry, opportunism and uncertainty (Coase, 1937; Williamson, 1985). While some literature sources question the effectiveness of contracts especially in view of uncertainties and opportunism, others overcome that by incorporating relational factors which facilitate information sharing, problem solving and adaptation, enhance mutual cooperation and solidarity necessary to minimise information asymmetries and opportunism (e.g., Cannon, Achrol and Gundlach; 2000) and overcome adaptation limits of contracts to uncertainties (Poppo and Zenger, 2002). This section defines relevant concepts for this chapter; (1) preference for contracts, (2) terms of contracts and (3) the contextual factors in which fishermen operate (Figure 4.1).

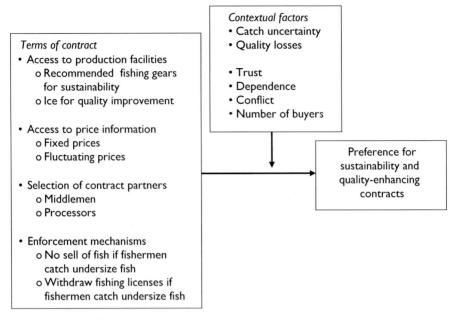

Figure 4.1. Conceptual framework.

4.2.1 Preference for contracts

The focus of this study is to determine fishermen's preference for contracts that oblige them to implement sustainable and quality–enhancing practices. Contracts are hereby understood simply as oral or written agreements between fishermen and buyers. Contracts outline the conditions that are offered in return to oblige fishermen to implement sustainable and quality-enhancing practices. In this research framework, fishermen would assess the extent to which they would prefer particular contracts within the context in which they operate. Preference for contracts is therefore defined as the extent to which a fisherman would be willing to sign up a particular contract. Preference will thus be mediated by contextual factors.

4.2.2 The terms of contract

One major contribution of the contract transactions among small-scale primary producers is that contracts address the market failures that producers face (Key and Runsten, 1999). As such, the terms of contracts between producers and buyers reflect constraints to be addressed and contractual obligation for both buyers and producers. In this study, the terms of contract are the conditions offered to fishermen in return for them to implement sustainable and quality - enhancing practices. These include; (1) access to production facilities; (2) access to price information; (3) selection of contract partner and (4) enforcement mechanisms. This chapter determines whether these conditions would influence preference for contracts.

Access to production facilities: In developing economies, inadequate use of appropriate production inputs and technologies has been attributed to poor markets and high costs for the technologies that many poor producers cannot afford (Key and Runsten, 1999). Given that this study deals with sustainability of the fisheries and quality of fresh fish, production facilities are considered as tools and facilities necessary to implement sustainable fishing practices (i.e., recommended fishing gears) and tools for quality improvement (e.g., ice). The chapter examines if access to these facilities would motivate fishermen to implement sustainable and quality-enhancing practices.

Access to price information: According to theory of perfect competition, balanced access to market information enhances balanced price negotiations (Coughlan *et al.*, 2001). Any imbalance in access to price information therefore skews bargaining advantage to those with price information. Although price information can be accessed in different ways such as media or marketing institutions, this study considers access to price information if fishermen have fixed prices in the contract which means that they know when prices will change.

Selection of contract partners: One of the strategic decisions for partners intending to engage in contracts is the selection of transaction partners. According to social- and network theory, past relationships between transaction partners may impede or enhance continuity of relationships or selection of new transaction partners (Wathne, Biong and Heide, 2001; Wuyts and Geyskens, 2005). In the setting of this study, selection of contract partner refers to selection between middlemen and large-scale processing factories both of which are buyers, but operate at different levels in the channel.

Enforcement mechanisms: From contractual relations perspective, enforcement is the disciplinary action taken when contract partners violate the terms of contracts (Antia and Frazier, 2001). In the setting of this study, i.e., common pool fisheries, sustainable practices tend to be enforced by public institutions to protect public interests. Hence enforcement mechanism is defined as the penalties against fishermen when they violate conditions for sustainable fishing. Specifically, the study considers withdrawing fishing license and no sell of fish if fishermen do not follow sustainable fishing, i.e., if they use bad gears (as evidenced by landing juvenile fish - i.e., less than 1.5kgs). Withdrawing fishing license implies that the enforcement is undertaken by public institutions that have the legal authority to revoke fishing licenses. No sell of fish means that the private sector, especially buyers, should enforce sustainable fishing by not buying fish for which unsustainable fishing is detected.

4.2.3 The context

The context defines the characteristics of the environment within which fishermen operate. We consider catch uncertainties and quality losses that directly affect the supply quantity and quality in the market; and trust, dependence, conflicts and number of buyers. We determine whether these factors may moderate fishermen's preference for the terms of contract.

Catch uncertainty: According to TCE (Williamson, 1985), uncertainties make it difficult for actors to predict future contingencies, specify outcomes and/or measure performance. Given that the setting of this study is a supply chain of fresh ecological products, we consider uncertainty of fish catch, i.e., production at source. We define catch uncertainty as the predictability of fish catch, which can be due to variation in fish stock, weather or seasonality.

Quality losses: Given that we are dealing with fresh products at small-scale primary producers' level, quality refers to the basic features of fresh products, i.e., freshness and texture of fish as quality criteria (Olafsdottir *et al.*, 2004). Degradation of freshness and texture may be aggravated by lack of cooling facilities (such as ice) in high temperatures, especially if fish takes long before landing, fishing and fish handling methods. Since freshness and texture are some of the most important quality attributes for fresh products, quality loss is reflected by the proportion of fish rejected by buyers at the point of sale either because fish is no longer fresh or has damaged fillet.

Trust: Trust reflects one's confidence, positive expectations and acknowledgment that a transaction partner will cooperate and advance mutual goals (Geyskens *et al.*, 1998). Trust influences attitude and the behaviour that actors display in working with their transaction partners. This study defines trust as the fisherman's belief that a middleman (buyer) is sincere in their transaction relationship.

Dependence: According to social exchange theory (Wrong, 1968), dependence creates power imbalance that enables independent partners to direct the activities of the dependent actors (Gundlach and Cadotte, 1994; Heide, 1994). Dependence has been conceptualised as the extent to which an actor requires specific resources from a partner and how an independent actor exercises control over the resources (Argyres and Liebeskind, 1999). Dependence when accompanied by uncertainties heightens governance concerns as independent actors may align transaction arrangements to serve their self interests (Heide and John, 1992; Williamson, 1985). Following Hewett and Bearden, (2001), this study defines dependence as the extent to which fishermen rely on middlemen to undertake their transaction relations.

Conflict: Conflicts arise when one party perceives that its interests are being opposed, impeded, or negatively affected by the activities of a transaction partner (Jap and Ganesan, 2000; Morgan and Hunt, 1994). When conflicts arise, actors may exit the relationship, voice their concerns to mutually resolve conflicts, or ignore the conflicts and remain loyal to the relationships (Ping, 1997). Such reactions depend on the quality of the relationship before the conflict, availability of alternative relationships, level of interdependence and criticality of the issues over which conflicts arise (Hibbard, Kumar and Stern, 2001). Thus conflicts may end or change or never affect the way transactions are governed. In this study, conflict refers to the degree of disagreements between a fisherman and a middleman over their transaction relationships.

Number of buyers: Number of buyers is viewed from the perspective of network ties. Existing literature discusses network relations from different perspective such as number and strength of ties (Granovetter, 1973) and network centrality (Anita and Frazier, 2001). Network ties influence information sharing necessary for the adaptation to volatile market situations (Achrol and Kotler, 1999). This study determines if the number of buyers a fisherman sells fish to may moderate preference for contracts.

4.3 Hypotheses

As envisaged in Figure 4.1, preference for contracts depends on the terms of contracts moderated by the context in which fishermen operate.

4.3.1 The effect of the terms of contracts

Access to production facilities: Appropriate production technologies are required to enhance sustainability as well as meeting quality standards. For example, use of recommended fishing gears is critical for sustainability because they protect juvenile fish from being caught. Similarly, specialised tools such as ice or cooling facilities are needed to ensure quality, i.e., freshness of fish. Access (or lack of it) to these production facilities may have different implications. On the one hand, fishermen need more fishing gears in order to increase the chances of catching fish in view of declining fish stocks. On the other hand, as a result of lack of cooling facilities, fishermen lose fish because of spoilage especially when they take long to land. In addition, as a result of lack of cooling facilities fishermen fail to bargain, say, for competitive prices for long for fear of losing more fish to spoilage. However, choosing ice faces practical problems. From the structural point of view, fishermen may require bigger boats (canoes) that can adequately carry ice, fishing gears and the fishing crew. This may entail further investments, for example, into large fishing boats or canoes that fishermen may need to undertake. Such investment may not be warranted without assurance of the sustainability of the fisheries. Therefore, fishermen may first need to protect the fisheries from further depletion to secure their business and more importantly, welfare before they can invest in quality improvement. We thus predict that:

H1: Fishermen will prefer contracts in which recommended fishing gears are provided over contracts that provide ice

Access to price information: Access to market information enhances market decisions. For the suppliers, price information may influence the decision about where to sell in case alternative competitive market outlets for the products exist, when to sell in case storage facilities for the product are available or when to produce in case production cycle can be controlled to take advantage of, say, competitive market conditions when they arise. This might be applicable to fishermen who have a very short production cycle, e.g., within a day. This means that when fishermen have price information, they may decide when to fish as well as how much to fish. By deciding how much and when to produce, fishermen may be able to control fishing effort that is critical for sustainability. Having price information, would also enhance fishermen's bargaining power and minimise uncertainty over price and ultimately, over income. One way to have price information is for fishermen to settle for fixed prices. Having fixed prices is also a common practice and motivation for producers, for example, in contract farming (FAO, 2001; Masakure and Henson, 2005). Therefore, in order to promote sustainable practices in a manner that minimises fishermen's price risks, we envisage that:

H2: Fishermen will prefer contracts in which fish prices are fixed over contracts in which prices are fluctuating

Selection of contract partner: Prior research provides considerable insights into the conditions that compel actors to select transaction partners such as social and structural bonds (Wuyts and Geyskens, 2005); history of good relationship quality (Hibbard, Kumar and Stern, 2001) and anticipation of better economic gains (Wathne, Biong and Heide, 2001). As a consequence, actors may prefer to continue with existing relationships or switch to competing ones. Anticipation of better economic gains may be an important motivation for small-scale fishermen because, being relatively poor, their welfare depends on their daily income (Platteau and Abraham, 1987). For instance, fishermen may intuitively expect better economic gains from direct transactions with processors because they are large scale buyers compared to small-scale middlemen. Anticipated economic benefits may be critical for fishermen who, because of declining fish stocks, tend to increase fishing effort to catch more fish to compensate for lost income. Hence, if fishermen can secure better economic benefits (e.g., better fish prices) when they catch less fish, they may be motivated to implement sustainable practices. By implication, implementing sustainable fishing practices means using recommended fishing gears that may lead to catching less fish. Therefore for fishermen to implement sustainable fishing practices in a manner that may enhance their welfare, they may select contracts partners from whom they expect better economic gains. To that extent, we envisage that:

H3: Fishermen will prefer contracts with processors over contracts with middlemen

Enforcement mechanisms: Contracts no matter how well-designed may still be violated in one way or another (Antia and Frazier, 2001) and that is why they often outline penalties for violations (Poppo and Zenger, 2002). Since the success of contracts depends on the context in which they are applied (FAO, 2001; Cannon, Achrol and Gundlach, 2000), property rights and the way sustainable practices are enforced in common pool fisheries might influence whether or not fishermen would be motivated to engage in contracts in the first place. For example, fishermen that may engage in contracts will be obliged to use good fishing gears and hence, they may catch less fish compared to those who may use bad gears outside contracts. Unless fishermen using bad gears outside contracts are punished, those under contractual obligation to use good gears might be losing. There might be different ways to punish those using bad fishing gears. One way would be to withdraw fishing gears and licences. This means that the public institutions that issue the fishing license should revoke them. Given the ineffectiveness of the public institutions and the enforcement costs involved in monitoring fishermen, this may not assure fishermen willing to engage in contracts that irresponsible ones would be dealt with. Another way to punish fishermen using bad fishing gears would be for buyers to refuse buying fish from fishermen using illegal gears. Such fishermen would be easy to identify because they would have juvenile fish and besides, fellow fishermen would also easily identify them. In that way, the private sector including the fishermen themselves would enforce sustainable fishing practices. Since it may be in the interest of fishermen to protect the fisheries as a fountain for their livelihoods and business, they may be motivated to engage in contracts where the private sector enforces sustainable fishing practices because the public institutions have so far been ineffective. We therefore predict that:

H4: Fishermen will prefer contracts in which buyers refuse to buy fish caught with bad gears over contracts where public institutions withdraw fishing gears and licenses.

4.3.2 The moderating effects of the contextual variables

Catch uncertainty: According to TCE, uncertainties no matter the form, create exchange hazards by requiring that parties adapt to unforeseeable circumstances (Williamson, 1985). Increasing catch uncertainty implies that fishermen run the risk of returning empty-handed from a fishing trip or with too little fish to be of any commercial value or meet livelihood needs. In the short term, fishermen may adapt to increasing catch uncertainty by investing in more fishing gears to increase their chances of catching fish. In the long-term however, catch uncertainty signals threats to sustainability and so to their livelihoods. In that case, fishermen may adapt to uncertainty by investing in recommended fishing gears to protect their future livelihood. In that case, investing in recommended fishing gears would not only increase chances of catching fish in the short-term but more importantly, secure the long-term sustainability of the fisheries. In view of the limited exit options for fishermen to enter into other sectors of the economy, we envisage that-

H5: The more fishermen face catch uncertainty, the more they will prefer contracts in which recommended fishing gears are provided over contracts that provide ice for quality improvement.

Quality losses: Given the non-availability of ice in landing sites coupled with high tropical temperatures, fishermen loose some fish by the time they land. Fishermen can minimise fish spoilage if they land and sell fish within a short period after catching or keep fish in ice in case they take long to land. Landing as soon as they catch fish may not always be practical, however. Due to declining fish stocks, fishermen take long to catch reasonable amount of fish. Sometimes they go long distances to the fishing grounds or face strong tides in the lake that make them fail to sail back to land in time (see Chapter 3). This leaves use of ice as a feasible option to minimise fish spoilage. We thus envisage that holding other things constant:

H6: Fishermen that incur more quality losses will prefer contracts in which ice is provided over contracts in which recommended fishing gears are provided

Trust: On the one hand, following the arguments that trust mitigates against opportunism (e.g. Morgan and Hunt, 1994), we envisage that fishermen may prefer to contract trustworthy middlemen. On the other hand, arguing that trustworthy middlemen pose no risk of opportunism and hence no need for contractual safeguards (e.g. Larson, 1992), one would expect that fishermen would prefer to contract untrustworthy partners in order to guard against their untrustworthiness. However, it might be in the interest of fishermen to be safeguarded by contracting middlemen who represent their interest. For example, when fishermen come back empty-handed from fishing trips, it may need a middleman who does not interpret that as opportunism, i.e., that a fisherman undertakes extra-contractual sell of fish. In that case, a fisherman may prefer contracts with middlemen already representing their interests rather than engaging with processors of whom they are not sure how they may react to problems beyond fishermen's control. In that sense, we hypothesize that:

H7: Fisherman that trust the middlemen will prefer contracts with middlemen over contracts with processors

Dependence: According to social theory (Wrong, 1968), fishermen who are dependent on middlemen might be vulnerable if middlemen align the terms of transactions to their advantage (Heide and John, 1992). However, existing literature demonstrates that small-scale primary producers engage in contracts with buyers they actually depend on for accessing necessary production inputs and services. For example, Platteau and Abraham (1987) note that small-scale fishermen enter into quasi-credit contracts with fish buyers to secure fishing gears as a way of insuring against the risk of falling into distress and income losses. Similarly, Masakure and Henson (2005) show that lack of access to competitive markets, market information and reliable sources of farm inputs compel small-scale farmers to engage in contracts with buyers who provide the inputs, guarantee markets and minimum prices for farm produce.

Problems with such interlocked markets and dependence are familiar, such as loss of bargaining power (Bardhan, 1980; Heide and John, 1998). One way to minimise such problems would be for fishermen to secure their own resources such as fishing gears and be free to decide where to sell the fish. However, abandoning dependency-based relationships may not always be a priority option for dependent actors. As Platteau and Abraham, (1987) note, a fisherman with low asset base and strong risk aversion may attach high importance to having a dependable source of subsistence credit before breaking a tie with a buyer (credit provider). Further, in a typical village society, dependence evolves around informal credit and gift exchanges that offer some insurance and social security (Platteau and Abraham, 1987). In that case, fishermen who are more dependent on middlemen may seek continuation of their relationships with middlemen rather than engaging with processors of whom they are not sure if they can guarantee social security in times of social stress. In view of these likely scenarios, we hypothesize that:

H8: a: Fishermen that are more dependent on middlemen will prefer contracts with middlemen over contracts with processors b: The more fishermen are dependent on middlemen, the more they will prefer contracts in which fishing gears are provided over contracts in which ice is provided.

Conflict: According to literature on channel relationships (e.g., Hibbard, Kumar and Stern, 2001), conflicts signal poor relationships and how actors react to them may have implications for the long-term viability and success of the relationship. Intuitively, fishermen would prefer to transact with buyers with whom they have good relationships – i.e., no conflicts. Hence one way to solve conflicts would be to engage in a new relationship. However, conflicts may not always end relationships because actors may adopt different ways to resolve conflicts depending on the nature and cause of the conflicts (Ping, 1997). In the relationships between fishermen and middlemen, unpredictable price changes resulting from lack of access to price information by fishermen is one of the major sources of conflicts. One way to minimise such conflicts is for fishermen and middlemen to have fixed prices as is often the case in contract farming (FAO, 2001). In absence of alternative means of obtaining price information, we envisage that fishermen who have a higher degree of conflicts with middlemen would prefer to have fixed prices.

H9: The higher the degree of conflicts fishermen have with middlemen, the more they will prefer, (a) contracts with processors over contracts with middlemen, and (b) contracts in which fish prices are fixed over contracts in which fish prices are fluctuating.

Number of buyers: Drawing from network theory (Granovetter, 1973), a fisherman who has many buyers might be in an advantage position in terms of accessing price information. The result of this could be that such fishermen may not necessarily have to develop stronger or contractual relations with particular middlemen to access price information. So if they prefer to contract, price information may not be an important incentive. So we hypothesize that:

H10: Fishermen that tend to sell fish to more buyers will prefer contracts in which prices are fluctuating over contracts in which fish prices are fixed.

The hypotheses are summarized in Table 4.1.

4.4 Methodology

This section discusses the data collection procedures, sample, measurements and validation. Data was collected through a survey questionnaire which consisted of multi-item scales measuring trust, dependence and conflict, catch uncertainty, demographic characteristics of the fishermen, other contextual variables plus a conjoint analysis task.

4.4.1 Data collection procedure
Sampling: In order to identify the sample for our study, we consulted the Kenya Marine and Fisheries Research Institute (KMFRI) for a sampling frame of fishermen. However, KMFRI

Table 4.1. Summary of hypotheses.

Terms of contracts		Main effects	Moderating effects					
Factor	Factor level		Catch uncertainty	Quality losses	Trust	Dependence	Conflict	Number of buyers
Access to production facilities	Recommended fishing gears	(+) H1	(+) H5			(+) H8b		
	Ice for quality management			(+) H6				
Access to price information	Fixed prices	(+) H2						
	Fluctuating prices						(+) H9b	(+) H10
Selection of contract buyer	Middlemen	(+) H3			(+) H7	(+) H8b		
	Processors						(+) H9a	
Enforcement mechanism	No sell of fish if fisherman catch undersize fish	(+) H4						

did not have any sampling frame and instead it had a list of fish landing sites. Consequently, fishermen were selected on the basis of these landing sites. However, the list of landing sites was outdated. For example, out of about 300 landing sites provided, the exact number of sites that were either operational or landing Nile perch was not known. The problem of having sampling frames with inadequate information has a long history (see Kish 1965; Sudman, 1976; Poate and Daplyn, 1993). Such problems can however be ignored or corrected (Kish, 1965). Although physical verification of the landing sites could have been the best solution, it was not feasible due to budgetary, time and accessibility constraints. For example, more than 20 beaches known to land Nile perch were in the islands and were left out due to safety reasons on water transport. In consultation with KMFRI researchers, we identified over 30 inland beaches that were known to be accessible by car.

Initial visits to some of these inland beaches revealed that in some beaches fishermen had either stopped landing Nile perch or relocated to other beaches leaving too few (as few as 2) fishermen landing Nile perch. We decided to target beaches with at least ten fishermen landing Nile perch. Although the number of fishermen was arbitrary, we avoided beaches with very few fishermen where tracing them could be time consuming without any assurance of finding them. In order to minimise coverage bias (Blair and Zinkhan, 2006), we finally selected 18 beaches covering 5 out of 8 districts of the Kenyan shore of Lake Victoria. We positioned ourselves at each beach the whole day waiting for - and interviewing - fishermen as they landed fish. In order to minimise selection bias (Blair and Zinkhan, 2006) that could arise from self selection or beach leaders favouring some fishermen to be interviewed, visits to the landing sites were unannounced. Upon landing, fishermen were approached and those who accepted were interviewed.

The sample: The sample consists of 278 fishermen. Out of the fishermen approached to be interviewed six did not participate for various reasons including lack of time after long fishing trips and unwillingness to be interviewed. The respondents were of ages ranging from less than 20 (7%) years to over 50 years (5%) with majority (78%) between 21 to 40 years while 10% between 40 to 50 years. About 4% had no formal education, about 65% had some years of primary education, about 30% had some years of secondary education and 2% had some tertiary education. Their fishing experience ranged from less than five years to over 20 years. About 51% of fishermen had, personally, no major additional income generating activity apart from fishing. About 18% had kinship relations (e.g. brother, sister, parents) with middlemen they were selling fish to at the time of the study.

The survey: To collect data, questionnaire interviews were conducted in English and two local languages (Luo, the language of one of the local fishing tribes and Swahili, the trade language for East Africa) that were commonly spoken among respondents. Three experienced research assistants from KMFRI helped in data collection. In order to ensure consistency in translation of the questionnaire, the research assistants were trained during which they translated the questionnaire from English to local language and back to English before the actual data collection. To determine the ease of implementing the survey, questionnaires were pre-tested twice in three beaches involving 11 fishermen. During pre-testing, fishermen were asked to comment on the length of the interview and the ease to understand the questions. Later research assistants were also asked to evaluate the ease to interpret and complete the questionnaire, and the ease with which respondents were able to follow the questions. Some modifications were done to the definitions and initial measurement scales. Initially, the measurement scales and conjoint profiles were to be rated on a 7 point Likert scale which proved difficult for the respondents to differentiate 2 from 3 and 5 from 6. This prompted us to change to a 5-point scale where 1 was strongly disagree and 5 was strongly agree. The use of 5-point Likert scale is common in empirical studies (e.g., Burnham, Frels and Mahajan, 2003; Cannon and Homburg, 2001; Joshi and Campbell, 2003).

Conjoint analysis task: In order to determine how fishermen may develop preferences for hypothetical contracts that oblige them to implement sustainability - and quality - enhancing practices, a conjoint analysis was used. Conjoint analysis is a multivariate statistical technique widely used to understand how consumers develop preferences for hypothetical products or services which are described by a number of attributes and attribute levels. It is based on the assumption that consumers are able to evaluate the value of a product or service by combining separate amounts of value provided by each attribute. As such, conjoint analysis is mainly used in consumer research (see Green, Krieger and Wind, 2001; Wittink, Vriens and Burhenne, 1994). However, it is also being applied to other aspects of marketing research such as evaluating channel relationships (Wathne, Biong and Heide, 2001; Wuyts *et al.*, 2004); or product positioning and design (Kaul and Rao, 1995).

To implement the conjoint tasks, a full profile presentation method often recommended for few (up to 10) factors was used (e.g. Green and Srinivasan, 1978, Hair *et al.*, 1998). However, one problem with 2^4 full factorial design, fishermen would have to evaluate 16 i.e. 2*2*2*2 profiles, excluding holdout profiles for assessing predictive validity. In order to minimise information overload and boredom, we used the fractional factorial main effect design (Hair *et al.*, 1998) to reduce the profiles to 8 while maintaining the orthogonality of the factors. In the end, fishermen evaluated 12 contracts which included four holdout profiles.

Setting the context for the conjoint task: To set the context within which the conjoint task was implemented, fishermen were first briefed about the objectives of the conjoint tasks, the attribute levels and how the profiles were to be evaluated. Fishermen were asked to consider the real life situation, i.e., the degradation of the Nile perch and the quality deterioration of the fish they catch and sell. Then, they were asked to imagine that they are being approached to engage in contracts that oblige them to use good fishing gears for Nile perch (i.e., gillnets of 12.7cm or hook number 5) as mandated in the Fisheries Act of Kenya (Kenya Government, 1991) and supply fresh non-bruised fish. Then conjoint profiles were presented and explained to the respondents, one at a time in a personal interview setting (see Box 4.1 for an example of two profiles). Respondents were asked to rate profiles on a Likert scale of 1 to 5, where 1 was least willing and 5, was most willing to sing up a particular contract.

Box 4.1 Examples of the conjoint profiles.

Profile 1
Question: To what extent would you be willing to sign this contract for you to use recommended fishing gears (12.7cm gill nets or hook number 5) and supply fresh non-bruised fish?

    You will be given ice on credit for keeping fish
    Fish prices will be changing anytime without notice
    Fishermen that catch fish of less than 1.5kg will not sell their fish
    This contract is with a processor

Least willing   1   2   3   4   5   Most willing

Profile 2
    You will be given recommended fishing gears on credit
    Fish prices will be agreed and periodically fixed in contracts
    Fishing license will be withdrawn if fishermen catch fish of less than 1.5kgs
    This contract is with middlemen

Least willing   1   2   3   4   5   Most willing

4.4.2 Measurements and validation

The multi-item measures measuring trust, conflict and dependency were adapted from existing literature (Jap and Ganesan, 2000; Andaleb, 1995) (see appendix 4.1 for the detailed measurement items). Catch uncertainty measured the predictability of fish catch relative to five years ago. Quality loss was measured as the proportion of fish rejected by buyers at the point of sale i.e., by the time fishermen land the fish. The number of buyers was measured as the number of middlemen the fishermen normally sells fish to. In beaches where fishermen sell fish in groups, the number of middlemen who usually buy fish from the group is used as a proxy for the number of buyers

In order to validate the multi-item measurements scales, i.e., trust, conflict and dependency, recommendations by Shook *et al.* (2004) to examine their unidimensionality, discriminant validity and reliability were followed. The measurements were examined for unidimensionality through exploratory factor analysis (Churchill, 1979). All items that loaded on multiple factors and/or with low loadings after varimax rotation (Hair *et al.*, 1998) were dropped. The retained items were imputed into a confirmatory factor analysis model in LISREL 8.72 (Jöreskog and Sörbom, 2005) with all factors. All factor loadings were significant ($t > 1.96$) (Byrne, 1998). The model fit statistics such as Comparative fit index (CFI) were above .95, Goodness of Fit Index (GFI) and Adjusted Goodness of Fit Index were both above .90 and Root Mean Square

Error of Approximation was ≤ .05. All these measures suggest good model fit (Schermelleh-Engel and Moosbrugger, 2003).

Further, the multi-item measurements were examined for discriminant validity by assessing pairs of constructs in a series of two - factor confirmatory models (Anderson, 1987; Bagozzi and Phillips, 1982). Each model was run twice, first, constraining the covariance between the two constructs and the variances to 1 and then removing the constraint. Following this, a Chi-square difference and changes in the Comparative Fit Index (CFI) (Byrne, 1998) were examined. For all models investigated, the Chi-square values were significantly lower for the unconstrained models than for the constrained models and the CFI values for the constrained models were lower suggesting poor fit. The changes in Chi-Square and CFI are given in Appendix 4.2. We further examined the discriminant validity following Fornell and Larcker (1981) who suggest that the average variance extracted from each item by the construct should be greater than the shared variance between the constructs. This was confirmed in all our models (see Appendix 4.3). The reliability of the measurement as measured by Cronbach alpha was very high.

Covariates: We also measured some demographic characteristics of fishermen including age, level of education, income, gear ownership, whether fishermen had other major sources and kinship relations. Age was measured in years. The level of education is the number of years of formal education. Income is measured as average monthly income. Gear ownership was measured as whether or not fishermen owned the fishing gears that they were using. Other major income source determines whether a fisherman personally has other major income sources besides fishing. Kinship relationship with buyers determines whether or not fishermen had biological relations with middlemen they were trading with, e.g., as brother, sister or parents. Tables 4.2 and 4.3 give the characteristic of the measurements and covariates and correlation matrix respectively.

4.4.3 The conjoint model
In order to validate the conjoint tasks, the data was examined for any peculiarities. One questionnaire had missing information on one profile and so it was dropped. Three other profiles had very extreme scores (e.g. similar scores across profiles) and they too were dropped. Then a conjoint analysis was undertaken on the remaining (274) respondents. To determine the validity of the model, first, we assessed the model fit. Then, we further assessed the predictive validity in predicting the holdout sample (Green and Srinivasan, 1990). In order to assess the model fit, we examined the Pearson correlation coefficients that give the correlation between the original and predicted preference scores. The Pearson coefficients (see Figure 4.2) revealed that the model did not adequately represent all respondents' data because some correlation coefficients were not significant.

A total of 8 questionnaires that did not show significant predictive validity were dropped leaving a sample of 266 respondents with an average Pearson correlation coefficient of .893 with standard deviation of .093.

Table 4.2. Summary of the purified constructs and other variables for fishermen' sample.

Variable	Operationalization	Number of items	Range	Mean	Standard deviation	Reliability
1 Catch uncertainty	Predictability of fish catch relative to five years ago.	1	1-5	4.37	0.96	
2 Quality losses	Proportion of fish spoilage by the time of landing		1-5	2.35	1.07	
3 Trust in middlemen	Fishers belief that main buyer is sincere and fulfils promises (Andaleeb, 1995)	4	1-5	3.68	1.18	0.872
4 Dependence on middlemen	Extent to which fishermen rely on middlemen in their business (Jap and Ganesan, 2000).	4	1-5	2.18	1.23	0.904
5 Conflict with middlemen	Degree of disagreements with middlemen (Jap and Ganesan, 2000).	2	1-5	2.18	1.22	0.795
6 Ownership of fishing gear	If a fisherman owns fishing gears and boats		0-1			
7 Number of buyers	Number of buyers the fisher normally sells fish to		1-7	1.9	1.4	
8 Average monthly income	Average monthly income (000 Kenya shillings)		5-90	14.55	15.12	
9 Other sources of income	If fishers has other major sources of income besides fishing and fish trading		0-1			
10 Kinship relations	If fisher has kinship relations with main buyer		0-1			
11 Age of fisher	Age in years		18-71	30.71	9.84	
12 Level of education	Years of formal education (in categories)		1-4	2.27	.547	

Table 4.3. Correlation matrix variables in fishers' sample.

Variables	1	2	3	4	5	6	7	8	9	10	11
1 Catch uncertainty	-										
2 Quality losses	.030										
3 Trust in middlemen	-.011	.084									
4 Dependence on middlemen	.030	-.151*	-.467**								
5 Conflict with middlemen	-.083	-.162*	-.547**	.448**							
6 Ownership over fishing gears	-.029	-.042	.060	-.016	.046						
7 Number of buyers	-.029	-.013	-.331**	.237**	.112	-.001					
8 Average monthly income	-.021	.088	-.072	-.095	-.143*	.239**	.088				
9 Other income sources	.060	-.014	.046	.073	.144*	-.037	-.155*	-.202**			
10 Kinship relations with middleman	.007	.059	.202**	-.185**	-.127*	.011	.035	-.050	.025		
11 Age of fishermen	.081	-.044	-.112	.019	.112	.167**	-.044	.058	.080	-.001	
12 Level of education	-.117	-.043	-.223**	.135*	.337**	.067	.075	-.084	.074	-.042	-.064

* Significant (p<0.05); ** Significant (p< 0.01) (all 2-tailed).

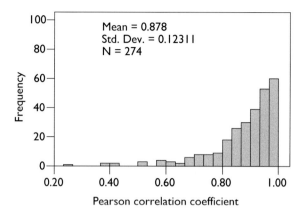

Figure 4.2. Pearson correlation coefficient for assessing model fit.

In order to determine the predictive validity for the model, we assessed the ability of the model to predict the preference for the holdout sample (Green, and Srinivasan, 1990). We examined the Kendall's tau for the holdout sample, which revealed that for 24 respondents the model did not generalise beyond the sample (see Figure 4.3). These respondents had low Kendall's tau of less than .30. They were also dropped leaving a final sample of 242 with an average Kendall's tau of .756 with a standard deviation of .168. The low predictive validity of the conjoint tasks could be attributed to possible inconsistencies in translation of the conjoint profiles given that not all fishermen were able to read and evaluate the tasks on their own.

4.4.4 Data analysis
After individually scrutinizing respondents, an aggregate analysis was performed with ordinary least squares regression analysis to test our hypotheses. The analysis was done in sequence; (1)

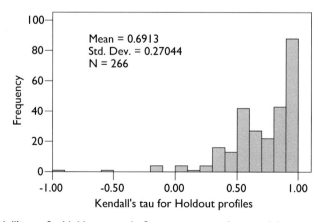

Figure 4.3. Kendall's tau for Holdout sample for assessing predictive validity.

the types of contracts fishermen prefer and (2) whether fishermen's preference for particular contracts was moderated by the contextual factors and 3) cluster analysis.

Determining the type of contracts fishermen prefer: In order to determine the type of contracts that fishermen prefer a regression analysis is used to explain preference for contracts as a function of the terms of contracts, i.e., main effects. In order to eliminate possible effect of fishermen's extreme preferences for particular profiles, preference scores for the eight calibration profiles were standardized to make them comparable across fishermen. The standardized preference scores were then used as the dependent variable. As the profiles were the unit of analysis, it means that with 8 calibration profiles, the total profiles used in the analysis was the product of the number of profiles and the number of respondents (i.e., 8 * 242 = 1936). After standardizing the preference scores, the attribute levels were coded following an effects coding scheme (Cohen and Cohen, 1983) in order to represent the different levels of the factors in the regression analysis. Under such a scheme, the first level of each factor (e.g., access to fishing gears) is coded as −1, and the other (e.g., access to ice) as +1. These attribute level dummies were then used as explanatory variables for the standardized preference scores.

Determining moderation effect: In order to determine whether fishermen's preference for contracts could have been moderated by the contextual variables, another regression analysis was undertaken. The moderating variables were defined by multiplicative products between the attribute level dummies and the relevant contextual factors. The fishermen's preference for contracts was estimated as a function of the terms of contract and their interaction with contextual variables. The change in R^2 was also tested for significance in order to determine if the model with moderating effects explained significantly higher variation in the preference for particular contracts than the main effects only.

To avoid the problem of multicollinearity, independent variables except for the attribute level dummies were mean-centred (Aiken and West, 1991) before the actual analysis. Nevertheless, to be certain about that, the variance inflation factors (VIF) and tolerance for both regression models were examined (e.g., Hair *et al.*, 2001). The VIF for all models were below 3 which is well below the cut-off point of 10 suggested in the literature (Hair *et al.*, 2001). Hence multicollinearity was not a problem in our analysis. In addition to checking for multicollinearity, an examination for heteroskedasticity was undertaken following Maddala (1992). The predicted values and standardized residuals for each model were examined for any relationship and none was observed. It was also concluded that heteroskedasticity was not a problem in our analysis.

Cluster analysis: In order to assess if there were identifiable segments of fishermen with similar preferences for particular terms of contract, cluster analysis was undertaken. Fishermen segmentation with respect to the types of contracts they prefer would be important for the implementation of the contracts. Cluster analysis was done on the partworths utilities of the standardised preferences in two steps: first, a hierarchical clustering procedure and then a K-means cluster analysis based on the results of the hierarchical cluster analysis.

4.5 Results

This section gives results of the analyses following the order in which they were undertaken, (1), the types of contracts fishermen prefer, (2) the moderating effects of the contextual variables and finally, (3): segmentation analysis

4.5.1 The types of contracts fishermen prefer

The type of contracts that fishermen prefer was determined by assessing the main effects of the terms of contracts on preference for contracts. Table 4.4 shows the results which show that the terms of contract explain about 25% of the variation in the fishermen's preference for contract. These results are consistent with the overall assumption of this study that preference for contracts will depend on the terms of contract. The results show that fishermen prefer contracts in which fishing gears are provided (p<.01) in contrast with those in which ice for quality management is provided. This is consistent with Hypothesis 1. The results also show that fishermen prefer contracts in which fish prices are fixed (p<.01) compared to those in which fish prices are fluctuating. These results are consistent with Hypothesis 2. Consistent with Hypothesis 3, the results also show that fishermen prefer contracts with processors (p<.01).

Furthermore, the results show that fishermen prefer contracts in which fishermen who violate conditions for sustainable fishing should not be allowed to sell fish compared to contracts in which fishing gears and licenses are withdrawn. These results are consistent with Hypothesis 4 (p<.01). This means that fishermen prefer that the private sector especially the buyers should be enforcing sustainable fishing by not buying fish from fishermen with bad gears. This is in

Table 4.4. Regression coefficients for the main effects of terms of contract on preference for contract (N = 1936 contracts).

Independent variables (factor level dummies)	Dependent variable: Standardized preference scores		
	Unstandardized coefficients	Hypothesis	Remarks
Access to fishing gears	.374 ***	(+) H1	Supported
Fixed price	.279 ***	(+) H2	Supported
Contract with processors	.053 ***	(+) H3	Supported
No sale for fish if sustainable practices violated	.045 ***	(+) H4	Supported
Statistics R^2 (Adj. R^2)	.255 (.253)		
F	139.52 ***		

*** Significant (p<.01) (one - tailed)

direct contrast with the status quo where the public institutions (the fisheries department in particular) confiscate or arrest fishermen with bad gears.

Although we did not have hypotheses to suggest which contract term is contributing most to fishermen's preference, the results show that access to production facilities contributed the most to the preference for contracts followed by access to price information, selection of contract partner and enforcement mechanisms in that order.

4.5.2 The moderating effects of the contextual variables

On average fishermen prefer contracts that provide them with fishing gears and price information, i.e., fixed prices. Fishermen also prefer to have contracts with processors and also that fishermen who use bad fishing gears should not be allowed to sell their fish. We assume that preference for such contracts could be influenced by the context in which fishermen operate. In order to test our assumption, a regression analysis was run where fishermen's preference for contracts was explained as a function of the terms of contract and the moderating variables. The model explains about 29% of the variation in the fishermen's preference for contracts. The difference in the R^2 between the main effects and interaction effects models is significant $\{F(8, 1617) = 9.186; p<.01\}$. Table 4.5 gives the results.

The results show that the more fishermen experience catch uncertainty, the more they will prefer contracts in which fishing gears are given. This is consistent with Hypothesis 5. The results also show that fishermen who incur higher quality losses prefer contracts in which fishing gears are given ($p<.05$). These results are inconsistent with Hypothesis 6 which predicted that fishermen that incur high quality losses would prefer contracts in which ice is given. This implies that fishermen prefer contracts in which fishing gears are given even though they incur more quality losses.

Contrary to Hypothesis 7 which predicted that the more fishermen trust middlemen, the more they would prefer contracts with middlemen, the results show that fishermen prefer contracts with processors ($p<.01$) even though they trust middlemen. Hypothesis 8a predicted that fishermen that are highly dependent on middlemen prefer contracts with middlemen. The results do not give sufficient evidence to support the prediction ($p>.10$). But the results show that the more dependent fishermen are on middlemen, the more they prefer contracts in which fishing gears are provided. These results are consistent with Hypothesis 8b ($p<.01$).

The results also show that the higher the degree of conflicts fishermen have with middlemen, the more they will prefer contracts with processors. Although there is only indicative evidence ($p<.10$), the results support Hypothesis 9a. In line with hypothesis 9b, the results further show that the higher the degree of conflicts fishermen have with middlemen, the more they prefer contracts in which fish prices are fixed ($p<.01$). Finally, the results did not support ($p>.10$) Hypothesis 10 which predicted that fishermen who tend to have more buyers prefer contracts in which prices are fluctuating.

Table 4.5. Regression coefficients for moderating effects for preference for contracts (N=1916 contracts).

Independent variables	Dependent variable: Standardized preference scores		
	Unstandardized coefficients	Hypothesis	Remarks
Terms of contracts			
Access to fishing gears	.606***	+ H1	Supported
Fixed prices	.284***	+ H2	Supported
Contract with processors	.055***	+ H3	Supported
No sale of fish without compliance	.045**	+ H4	Supported
Moderating variables			
Access to fishing gears * catch uncertainty	.081***	(+) H5	Supported
Access to ice * quality losses	-.029**	(+) H6	Rejected
Contract with middlemen * trust in middlemen	-.054***	(+) H7	Rejected
Contract with middlemen * dependence on middlemen	.015	(+) H8a	Not supported
Access to fishing gears * dependence on middlemen	.071***	(+) H8b	Supported
Contract with processors * conflict with middlemen	.025*	(+) H9a	Supported
Fixed price * conflict with middlemen	.097***	(+) H9b	Supported
Fluctuating price * number of buyers	.012	(+) H10	Not supported
Statistics R^2 (Adj. R^2)	.288 (.284)		
F	54.56***		

*** Significant (p<.01); ** significant (p<.05) *significant (p<.10) (one - tailed)

4.5.3 Determining possible fishermen's segments

The preceding results show that fishermen facing different contextual variables prefer different contracts. We assume that there could be homogeneous segments of fishermen with respect to their preferences for the particular types of contracts. To determine this, we first undertook a hierarchical cluster analysis with Ward's method to get an idea about the possible number of distinct groups with similar preferences using standardised data. Following Hair *et al.* (1998), we inspected the agglomeration coefficients for large changes in the coefficients which suggest that very dissimilar clusters have been combined. Large changes in the coefficients occurred when moving from 5 to 4 clusters, from 3 to 2 clusters, and from 2 to 1 cluster, suggesting either five-; three-; or two- homogeneous clusters.

Then, we ran K-means cluster analysis with five; three and two clusters, in which we used the cluster means from hierarchical cluster analysis to start the K-means cluster analysis. Table 4.6 gives the mean utilities for each of the clusters from the K-means cluster analyses. A comparison of the clusters reveals similarities and differences in preferences for different contracts. But

Table 4.6. Mean (SD) utilities for clusters from K-means cluster analysis (N = 242 fishermen).

Factor	Production facilities	Selection of contract partner	Price information	Enforcement mechanisms	Cluster size (N, %)
Factor level	**Recommended fishing gears**	**Processor**	**Fixed price**	**No sell of fish without compliance with sustainability**	
Two-cluster solution					
1	.606 (.254)	.061 (.234) [a]	.342 (.338)	.060 (.237) *	174 (71.90)
2	-.243 (.334)	.033 (.449) [a]	.133 (.488)	-.008 (.308) *	68 (28.10)
Test statistics	F= 448.70; df = 1; <.01	F= .422; df = 1; p>.10	F= 14.29; df = 1; p<.01	F= 3.306; df = 1; p<.10	
Three cluster solution					
1	.543 (.266) [b]	.066 (.242) [c]	.485 (.220)	.043 (.234) [d]	141 (58.26)
2	-.408 (.286)	-.013 (.444) [c]	.277 (.373)	-.044 (.269) [d y]	45 (18.60)
3	.549 (.363) [b]	.076 (.328) [c]	-.221 (.301)	.104 (.293) [d x]	56 (23.14)
Test statistics	F= 191.63; df = 2, p<.01	F= 1.292; df = 2; p>.10	F= 133.41; df = 2; <.01	F= 4.172; df = 2; p<.05	
Five cluster solution					
1	-.253 (.348)	-.477 (.308)	-.182 (.277) [i]	-.101 (.401) [j]	12 (5.96)
2	-.552 (.215)	.008 (.417) [f]	.379 (.239)	-.007 (.207) [j]	24 (9.92)
3	.749 (.161)	-.005 (.186) [f (g)]	.289 (.202)	.057 (.216) [j]	114 (47.11)
4	-.231 (.305) [e]	.317 (.345)	-.351 (.317) [i]	.049 (.342) [j]	33 (13.64)
5	.208 (.189) [e]	.144 (.252) [f(h)]	.681 (.205)	.053 (.268) [j]	59 (24.38)
Test statistics	F= 249.24; df = 4; p<.01	F= 23.760; df = 4; p<.01	F= 121.04; df = 4; <.01	F= 1.246; df = 4; p>.10	

Note: Means bearing the same superscript are not significantly different (p>.10), for example, [a] = implies that mean utilities for contract with processors for cluster 1 and 2 are not significantly different; Means with different or no superscript are significantly different (p<.05);
* Means significantly different at 10%

the five-cluster solution reveals more between group heterogeneity than the other cluster solutions. We therefore think that the five-cluster solution could segment fishermen better. We briefly describe the clusters in the five cluster solution.

Cluster 1 which is the smallest among all clusters, is very unique. It is the only cluster that has a high utility for contracts with middlemen. We call this cluster a *Clan* because middlemen belong to the same local communities as fishermen. This may compel some fishermen to continue trading with the middlemen.

Cluster 2 is similar to cluster two in the three- and two-cluster solutions. This cluster primarily prefers contracts in which ice for fish quality management is provided and in which fish prices are fixed. This cluster has the highest preference for contracts in which ice for quality management is provided among all clusters in all cluster solutions. We call this cluster *Quality sensitive*.

Cluster 3 comprises fishermen that primarily prefer contracts in which fishing gears are provided and also in which fish prices are fixed. It is the largest cluster and also the one that has the highest preference for contracts in which fishing gears are given among all clusters in all cluster solutions. Given their strongest preference for contracts in which recommended fishing gears are provided, we call this cluster *Green* fishermen that seek sustainability above anything else.

Cluster 4 comprises fishermen that cannot be uniquely identified in their preference for particular contracts. In general, they prefer contracts in which fish prices are fluctuating, contracts with processors and ice is provided, in that order. Due to lack of decisive preference for a particular term of contracts, we call this cluster as *Opportunists* who may want to grab anything that comes their way.

Finally, *Cluster 5* comprises fishermen who primarily prefer contracts in which fish prices are fixed and, to a much lesser extent, they also prefer contracts in which fishing gears are provided. In view of their preference for contracts in which fish prices are fixed that minimises price risks for the fishermen, we call this cluster a *Price risk-averse* cluster.

In order to identify fishermen that prefer particular contracts, we profiled the segments according to their background and demographic characteristics. To do that, we used the Tukey's honestly significant difference test to compare cluster means for age, level of education, catch uncertainty, quality losses, trust, dependence, conflict and number of buyers. In order to determine if there was any association between demographic characteristic that were measured as ordinal variables such as kinship relations, ownership of fishing gears and having other major income generating activities besides fish trading with cluster membership, we used Chi-Square statistics.

For the demographic characteristics, the results show significant difference in age and a significant association between kinship relations and having other income activities with cluster membership (p<.05) (see Table 4.7) and in the level of trust and dependence (p<.01) and for conflict and number of buyers (p<.10) among clusters (see Table 4.8).

Cluster 1- the Clan has the highest proportion (33%) of fishermen that have kinship relations (e.g. brother, sister, parents or direct cousin) with middlemen. This may suggest that the Clan prefers contracts with middlemen may be due to their kinship relations. Cluster 2 – the Quality sensitive fishermen are on average the youngest (27 years) compared to the rest of the fishermen. In addition, although this cluster highly trusts middlemen, it is not highly dependent on middlemen. In fact it is the least dependent on middlemen among all clusters.

Cluster 3 – the Green fishermen have the highest proportion of fishermen (about 63%) that had no other income generating activities besides fishing. This may explain why they prefer to have recommended fishing gears not only to increase the chances of catching fish but most importantly to promote sustainable fishing to protect their only major source of livelihood. The Green fishermen also have the lowest proportion (about 11%) of fishermen with kinship relations with middlemen. Although this cluster has the highest number of buyers among all clusters, it has the lowest level of trust in middlemen.

Cluster 4 – the Opportunists have relatively high trust and dependence in middlemen but lower conflict. Like in the preferences for the terms of contracts, this cluster cannot be uniquely profiled by their demographic or background characteristics. Cluster 5 – the Price risk averse has the highest dependence on - and conflict - with middlemen among all clusters. This might explain why they primarily prefer contracts that provide price information – the main source of conflict.

In short, in spite of the fact that some differences in background and demographic characteristics are moderate, the differences in characteristics of the segments explain the differences in the preferences for particular contracts. For some characteristics such as catch uncertainty, it might be understandable that it does not characterise any particular segment because it cuts across all fishermen and it is beyond any fisherman's control.

4.6 Discussion and implications

The objective of this chapter was to determine factors that would influence fishermen to engage in contracts that oblige them to implement sustainable and quality-enhancing practices. Fishermen prefer contracts that provide *recommended fishing gears* over those that provide *ice* for quality improvement. Fishermen that prefer contracts in which recommended fishing gears are provided are particularly those that face higher *catch uncertainty*. Increasing uncertainty means that such fishermen are not able to catch reasonable amount of fish for their business and more importantly, for their daily livelihood. On that note, access to fishing gears might

Table 4.7. Demographic characteristics for the five clusters (242 fishermen).

Cluster	Age	Level of education	Monthly income (000 KSh)	Kinship relations with middlemen Yes (n, %)	No (n, %)	Ownership of fishing gear Yes (n, %)	No (n, %)	Other income generating activity Yes (n, %)	No (n, %)	Cluster name
1	35.55	2.09	12.573	4 (33.3)	8 (66.7)	7 (58.3)	5 (41.7)	6 (50.0)	6 (50.0)	Clan
2	27.05^a	2.33	10.071	5 (20.8)	19 (79.2)	15 (62.5)	9 (37.5)	13 (54.2)	11 (45.8)	Quality sensitive
3	30.59	2.34	15.732	12 (10.5)	102 (89.5)	81 (71.7)	32 (28.3)	42 (37.2)	71 (62.8)	Green fishermen
4	27.84^b	2.19	12.390	7 (21.2)	26 (78.8)	23 (69.7)	10 (30.3)	20 (60.6)	13 (39.4)	Opportunists
5	33.51^{ab}	2.22	11.878	16 (27.6)	42 (72.4)	44 (74.6)	15 (25.5)	37 (62.7)	22 (37.3)	Price risk-averse
Test statistics	F= 3.31; df = 4; p<.05	F= 1.03; df = 4; p>.10	F= 1.77; df = 4; p>.10	χ^2 = 10.08; df = 4; p<.05		χ^2 = 2.15; df = 4; p>.10		χ^2 = 12.81; df = 4; p<.05		

Note: The values for education are categorical where 1 equal no formal education and 4 some years of tertiary education.
Means bearing same superscript are significantly different (p<.10)

Table 4.8. Contextual characteristics for the five clusters (n = 242 fishermen).

Cluster	Catch uncertainty	Quality loss	Trust	Dependence	Conflict	Number of buyers	Cluster name
1	4.55	15.91	3.93	2.50	3.18	2.18	Clan
2	4.62	8.33	4.56^i	1.87^{kh}	2.90	1.67	Quality sensitive
3	4.27	13.07	3.20^{ij}	3.03^k	3.19	2.13^{m*}	Green fishermen
4	4.71	12.10	4.27^i	2.81	2.90^{r*}	1.90	Opportunists
5	4.49	12.63	3.92^i	3.16^h	3.64^{r*}	1.53^{m*}	Price risk-averse
Test statistics	F= 2.28 df = 4; p>.10	F= .88 df = 4; p>.10	F= 12.07 df = 4; p<.01	F= 3.93 df = 4; p<.01	F= 2.21 df = 4; p<.10	F= 2.14 df = 4; p<.10	

Note: Means bearing same superscript are significantly different (p<.05);
* Means significantly different (p<.10)

mean that in the short-term fishermen may increase the chances of catching fish, i.e., if they acquire additional gears.

Another important aspect of this result is that by accessing recommended fishing gears, such fishermen seek to promote sustainability of the fisheries which is very crucial for their livelihoods and that of the local communities in terms of food security and nutrition. Perhaps, what is more interesting about these results is that by virtue of wanting recommended gears, fishermen would in the short-run catch less fish because the juveniles would no longer be caught – at least not as much as they would be caught with bad gears. It is therefore intriguing to note that fishermen are willing and enthusiastic about access to recommended gears despite the possible short-term business and welfare implications. It is in fact more interesting to note that fishermen prefer gears regardless of the amount of fish spoilage they face. Improving quality has immediate short-term economic benefits yet fishermen seem to be foresighted by wishing to protect the fisheries first before improving quality. This is the case perhaps because fishermen are already experiencing the consequences of declining fish catch.

Another group of fishermen who prefer contracts in which fishing gears are provided are those that are more *dependent* on middlemen. A number of fishermen depend on middlemen for informal loans either in cash or in kind in form of fishing gears and equipment. When these fishermen obtain loans, they tend to be obliged to repay the loan in form of fish supply. Consequently, they become tied up without any chance to sell fish to other buyers. Perhaps, the most exploiting part of such loans is that the period of repayment and amount of repayment are never clear. To that extent, fishermen who obtain such loans from middlemen pay for them for as long as the fishing gears are operational. Therefore, having own fishing gears would free the fishermen from such interlocked markets and holding other things constant, enhance their bargaining power and decision about where to sell their products.

This study also shows that fishermen in general and in particular those that tend to have a higher degree of *conflicts* with middlemen prefer contracts in which fish prices are fixed. Abrupt changes in fish price at landing sites tend to be the major source of conflict between fishermen and middlemen. What makes matters worse for fishermen is that middlemen take advantage of the lack of cooling and storage facilities that limit the ability of fishermen to bargain for long for fear of losing most of their fish or to keep fish in the event of very low prices (see Chapter 3).

Another important result from this study is that fishermen generally prefer *contracts with processors* which could be seen in the context of expected economic gains and social relations with middlemen. For example, fishermen that have higher degree of *conflicts* with middlemen are more inclined for contracts with processors. The importance of expected economic benefits cannot be overlooked, of course. This can be deduced from the fact that even fishermen that trust the middlemen prefer contracts with processors. Although trust has previously been found to be a switching barrier for actors to engage new transaction partners (e.g. Wathne,

Heide and Biong, 2001), this study shows that fishermen regardless of their trust in middlemen prefer to switch to processors. Fishermen may implicitly expect better economic gains when dealing with large-scale processors than middlemen who, after-all, benefit from the same processors. Expected economic benefits have previously been noted to influence switching to new transaction partners (Wathne, Heide and Biong, 2001). Moreover contracts with processing factories would bring fishermen closer to the international channel unlike in their current position.

The results also show that fishermen prefer contracts when sustainable fishing can be enforced on other fishermen outside the contacts as well. Although contracts no matter how well crafted may still need to be enforced, sustainable fishing especially in the common pool resources also needs to be enforced. Generally, enforcement of sustainable fishing is undertaken by the public institutions such as the fisheries department. According to fishermen, such institutions are not only ineffective but also tend to be biased against them in their approach. The results of this study show that fishermen generally prefer contracts when those that violate conditions for sustainable fishing do not sell their fish. The implication of this result is that private institutions including buyers, fishermen groups and local institutions may have to take responsibility to ensure that responsible fishermen have better rewards. It is worthwhile to note that some fishermen groups and local institutions in some beaches already use social control to enforce sustainable fishing methods (see chapter 3).

The implication of preferences for different contracts is that no single contract can be applied. However fishermen segments cannot be identified with particular physical location such as landing sites. Whereas this may render locating fishermen for particular contracts difficult, one way to overcome that is to follow self selection. This means that different contracts could be crafted and be brought to the different landing sites where fishermen would then self select the contracts that best suit their preferences.

In conclusion, this study has established that fishermen are enthusiastic about the contracts for sustainability and quality. Although the proportion of quality losses fishermen incur due to lack of quality management facilities and poor handling did not enhance the preference for contracts in which ice is provided, it may not necessarily mean that fishermen do not need ice. It could be that the choice between ice and fishing gears was a difficult one to undertake. As one fishermen said during the interviews, "*You cannot fish without fishing gears, but you can still have fresh fish without ice.*" Hence fishing gears and tools for quality improvement should not be given as a choice between them but rather as complementary if both sustainability and quality have to be concurrently improved among fishermen.

4.7 Study limitations and future research

Despite the strong support for our conceptual framework, there are some limitations that present opportunities for further research. This study is limited to fishermen that at least catch

Nile perch, the export fish from Kenya. Although restricting our sample to such fishermen was convenient to study the export channel, a general study to include all fishermen regardless of the fish type they catch and the output markets they target would be important because sustainability of all fisheries of Lake Victoria is needed.

This study considered only four terms of contracts. Although this was good for the fishermen to evaluate a relatively small number of profiles, increasing the number of terms of contracts may have given a wider spectrum for the fishermen to choose from. Future research should therefore consider including more terms of contracts such as period of contracts, mode of payment; type of contract, i.e., group or individual and also testing for possible interactions. Although including more factors may increase the number of profiles to evaluate, future research may consider using other conjoint methods such as the trade off that may handle increased number of factors.

Future research may also consider separating the choice between fishing gears and tools for quality management which were given as a trade off in this study. Separating them may give better insights in the way fishermen consider their necessity in stimulating their preferences for sustainability and quality-enhancing contracts.

Appendices

Appendix 4.1. Construct items, path coefficient and T. values

In this appendix we provide information for measurement including: factor loadings, T-values, eigenvalues and Cronbach Alpha.

Trust (Alpha = 0.872, eigenvalue = 2.910)	Factor loadings	T-values
This buyer is very dependable	0.75	13.88
This buyer is sincere	0.86	15.88
This buyer has good reputation	0.78	13.62
This buyer always fulfils his promise	0.79	14.55
My relationship with this middleman is satisfactory	Dropped	

Adapted from Andaleeb, (1995)

Conflict (Alpha = 0.795; eigenvalue = 1.98)	Factor loadings	T-values
The relationship between me and this middleman is very good (R)	0.97	18.95
We have significant disagreements in our business relationship	0.63	10.24
We frequently quarrel with this middleman on issues of our business	Dropped	

Adapted from Jap and Ganesan, (2000)

Dependence (Alpha = 0.904; eigenvalue = 3.11)	Factor loadings	T-values
I can easily find other middlemen to buy my fish (R)	0.91	17.79
I fully depend on this middleman for my business	0.81	9.89
I can easily find a better middleman than the current one (R)	0.86	16.28
I can easily sell fish even if this middleman stops buying my fish(R)	0.79	14.75

Adapted from Jap and Ganesan, (2000)

Appendix 4.2. Test results for Discriminant Validity of constructs

Chi-Square (and CFI) differences between constrained and unconstrained models [a]

Fishers sample	1	2
1 Trust on middleman	-	
2 Dependence on middleman	381.17[b](-0.26)	
3 Conflict with middleman	117.01 (-0.11)	89.396(-0.15)

[a] The critical value ($\Delta X^2 > 3.84$) was exceeded in all tests

[b] Should read as X^2 of the constrained model including trust and dependence was 381.17 higher than the X^2 of the same model when unconstrained. CFI of the constrained model was 0.26 lower than the CFI of the free model.

Appendix 4.3. Discriminant Validity tests following Fornell and Larcker (1981)

Construct	1	2	3
1 Trust on middlemen	.636[a]		
2 Dependence on middlemen	.260	.711	
3 Conflict with middlemen	.563	.221	.706

[a] The diagonal gives the average variance extracted by the construct and the off diagonal gives the shared variance between constructs. In all the constructs the variance extracted by the constructs from each item is greater than corresponding shared variance.

Chapter 5

Factors influencing preference for sustainability and quality-enhancing contracts among middlemen

5.1 Introduction

This chapter investigates the factors that may influence middlemen's preferences for contracts that oblige them to promote sustainability of the fisheries and quality- enhancing practices in their transactions. To do that, the chapter translates the theoretical framework from Chapter 2 and applies it to the middlemen's setting. Middlemen are the intermediate buyers of fish from fishermen and supply to large-scale processing factories. Extending the theoretical framework to middlemen is necessary on two grounds. First, middlemen have a double loyalty, i.e., to fishermen and processing factories which, according to literature on market orientation may influence their buying and supplying behaviour. For example, market orientation influences customer-oriented behaviour of suppliers and supplier-oriented behaviour of buyers (e.g., Langerak, 2001). Second, extending the analysis to middlemen increases the feasibility of contracts to coordinate the activities between fishermen and middlemen in implementing sustainable and quality-enhancing practices. Although economics literature suggests a number of factors that motivate buyers and small-scale primary producers to engage in contracts (e.g., Key and Runsten, 1999; FAO, 2001), it is not known what factors would influence middlemen to engage in contracts that oblige them to promote sustainable and quality-enhancing practices in their buying and supplying transactions. This chapter uses conjoint analysis.

This chapter is organized as follows; first, the relevant concepts are defined followed by hypotheses. Then, we discuss the research methodology outlining the data collection procedures, measurements and validation. Results and discussion conclude the chapter.

5.2 Concepts

Building on transaction costs economics (TCE) and agency theory contracts are designed and chosen to match known exchange hazards created by free market failures such as specialized asset investments, opportunism, information asymmetry and uncertainty (Williamson, 1985). The success of contracts depends of the prevailing context in which they are applied (Cannon, Achrol and Gundlach, 2000) such as availability of profitable markets, the physical-, social- and/or regulatory environments (FAO, 2001). Although the necessity and impact of these factors may be different on the demand and supply sides of the middlemen, they may influence the types of contracts middlemen prefer with their suppliers and with customers. The underlying assumption in this study is that middlemen's preference for sustainability- and quality-enhancing contracts will depend on, (1) the terms of contract, and (2) contextual factors in which middlemen operate (Figure 5.1).

5.2.1 Preference for sustainability and quality-enhancing contracts

The focus of this study is middlemen's preference for contracts that oblige them to promote sustainability and quality-enhancing practices. Middlemen's preference for contracts is therefore defined as the extent to which middlemen would be willing to sign up a particular contract that oblige them to buy and supply good quality fish that is sustainably caught. Quality is defined in terms of freshness and texture such that good quality fish is fresh and also not crushed. Sustainably caught fish is fish that is caught by recommended fishing gears. For Nile perch, recommended fishing gears are supposed to catch fish not less than 1.5kgs. Fish of less than 1.5kg is considered a juvenile (Kenya Government, 1991) which can be caught by bad fishing gears. Hence contracts oblige middlemen to buy and supply fresh fish of not less than 1.5kgs.

5.2.2 The terms of contract with fishermen and processors

The terms of contract are the conditions under which middlemen are asked to engage in sustainability and quality-enhancing contracts with fishermen and processors. These conditions include in the purchase side, (1) price information, (2) period of contract; and (3) selection of suppliers and, in the supply side, (1) price information; (2) period of contract and

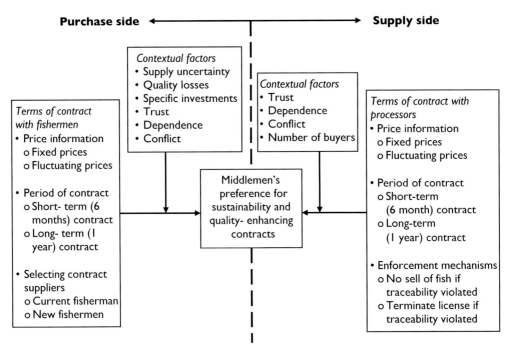

Figure 5.1. Conceptual framework.

(3) enforcement mechanisms. We will determine how the terms of contracts may influence middlemen's preference for contracts. In view of the similarities of the terms of contracts in the supply and demand sides, we will define them together.

Price information from and for middlemen: According to TCE, information asymmetry is one of the factors that necessitate contractual governance. In line with this, literature on contracts between buyers and small-scale primary producers demonstrate that access to market information is one of the motivating factors for the primary producers to engage in contracts to minimize price risks (Masakure and Henson, 2005; Key and Runsten, 1999). This study considers price information in the context of middlemen giving price information to fishermen and also middlemen having price information from processors.

Price information from middlemen varies between fixed prices and fluctuating prices. *Fixed prices* imply that middlemen agree on price with fishermen which remain fixed for a certain period. *Fluctuating price* means that middlemen can change the prices they offer to fishermen anytime without informing them of the pending price changes.

Similarly, price information for middlemen varies between fixed and fluctuating prices. *Fixed factory prices* imply that middlemen agree on price with processing factories that will periodically remain fixed. *Fluctuating factory price* means that processing factories can change (increase or reduce) prices anytime without informing middlemen of pending changes.

Period of contract with fishermen and, with processors: Contracts by nature lock small-scale producers into transaction with buyers without chance to try other buyers (Key and Runsten, 1999). Similarly, buyers may be locked up without possibility to try other suppliers. Hence the period within which buyers and producers are locked into contract relationship may have both costs and benefits. For example, long-term contracts means that actors may face higher opportunity costs in terms of possible competitive conditions that may arise from other buyers or suppliers. In the context of this study, period of contract refers to the time frame that middlemen will be obliged to trade with contracted fishermen or processors without the possibility to switch to new contract partners.

Period of contract with fishermen is the period in which middlemen and fishermen cannot terminate the contract. It varies between short-term contract (i.e., 6 months) and long term contract (1 year). The short-term contracts mean that middlemen and fishermen will be able to terminate contract only after 6 months. The long-term contract means that middlemen and fishermen will be able to terminate contract only after 1 year.

Period of contract with processors is the period which middlemen (and processing factories) cannot terminate the contracts. Similarly, it varies between short-term contract (i.e., 6 months) and long term contract (1 year). The short-term contracts mean that middlemen and processors will only terminate contract after 6 months. The long-term contract means that middlemen and processing factories will only be able to terminate contract after 1 year.

Selecting contract supplier: According to social and network theory (Granovetter, 1973), actors share close prior ties with some transaction partners while they have loose or no prior ties with other actors. The strength of such ties becomes important when actors are faced with decisions to select transaction partners for long-term or new transactions altogether (Wuyts and Geyskens, 2005). In the setting of this study, selecting suppliers refers to middlemen selecting between the fishermen they currently trade with and new fishermen.

Enforcement mechanism: Enforcement of the terms of contracts depends to some extent on whether or not an enforcement mechanism exists that penalizes breach of contract (Fafchamps, 1996; 2004). However, imposing penalties may also depend on whether the behaviour of the agents can easily be monitored. To overcome that, contracts tend to outline roles, obligations as well as mechanisms and penalties for non-compliance. This study considers traceability as an enforcement mechanism. With traceability, middlemen are obliged to identify fishermen from whom they buy fish. It varies between terminating traders' license and no sell of fish if fish traceability to fishermen is violated. Terminating license means that middlemen will lose their trader's license and are not allowed to buy and supply fish anymore. No sell of fish means that middlemen, who cannot trace the source of fish when they buy and supply undersized fish, will maintain their licenses and supply contracts but they will not sell any fish that is not traceable when juvenile fish is detected.

5.2.3 The contextual factors

The context defines the characteristics of the environment within which middlemen transact with fishermen and with processors. Middlemen may face different conditions in the input and output markets. This study considers contextual factors in the input markets, i.e., supply uncertainty, specific investments, trust, dependence and conflicts. In the output markets, we consider quality losses, trust, dependence, conflicts and number of buyers. In view of the similarities in the theoretical reasoning for the relational factors, we will define them together.

Supply uncertainty: According to TCE, uncertainties may result from external factors such as volatility in market place (Williamson, 1985) arising from unpredictable consumer demands or competitor actions (Joshi and Stump, 2003). Whereas, uncertainties over fish supply may arise due to ecological factors such as seasonality of fish production, middlemen may face supply uncertainty due to inability to observe fisherman - for instance, if fishermen engage in extra-contractual sale of fish. Inability to monitor an agent's behaviour is a major source of uncertainty in principal-agency relationships (Eisenhardt, 1989b). In the context of this study, supply uncertainty refers to the extent to which middlemen are able to predict fish supply. Although it would be logical to consider demand uncertainty in the output markets, demand for Nile perch fish is assumed to be certain because the processing factories are already not operating at full capacity due to inadequate fish supply.

Specific investment: According to TCE, specific investments raise concerns for the investor in that rents from such investments may be expropriated by self-interested partners (Lusch and Brown, 1996). As such, specific investments motivate investors to engage in contracts to safeguard their investments (Williamson, 1985). This study considers specific investments as fishing gears and equipment that middlemen give to fishermen to secure fish supply.

Quality losses: At the middlemen's stage, quality losses may occur due to mishandling of fish when packing, poor transportation (in bad roads) and when they take long to deliver fish to the processing factories without proper storage facilities (Olafsdottir *et al.*, 2004). In this study, quality loss is the proportion of fish rejected at the factory due to quality defects.

Trust in fishermen and, in processors: *Trust* is frequently described as the willingness to rely on an exchange partner in whom one has confidence (Moorman, Zaltman and Deshpande, 1992). An important aspect of this definition is the notion of trust as a belief or an expectation about an exchange partner that results from the partner's positive intentions and reliability (Ganesan, 1994). It has been widely documented that trust minimizes opportunism and advances mutual interests (e.g., Brown, Dev and Lee, 2000; Cannon, Achrol and Gundlach, 2000). This study considers middlemen's *trust in fishermen* and in processors. Trust in fishermen is the extent to which middlemen believe that fishermen are honest, reliable and sincere in their transactions with the middlemen. Similarly, *trust in processors* is the extent to which middlemen believe that processors are honest, reliable and sincere in their transactions with the middlemen.

Dependence on fishermen and, on processors: According to Hewett and Bearden (2001), dependence on a transaction partner can be defined as the extent to which an actor relies on the relationship to accomplish important goals, or adheres to a partner's specific requests. Dependent actors may take directives from independent partners who may align the governance arrangements to their advantage (Gundlach and Cadotte, 1994; Heide and John, 1992). This study defines middlemen's *dependence on fishermen* as the extent to which a middleman relies on fishermen to fulfil their transaction activities. Likewise, middlemen's *dependence on processing factories* is defined as the extent to which middlemen rely on processors to fulfil their transaction activities.

Conflicts with fishermen and with processors: According to Hibbard, Kumar and Stern (2001), one party may engage in actions that a transaction partner may in one way or another consider destructive for their relationship. The cause of conflicts and the partners' reactions may shape the future of transaction relationship (Ping, 1997). The current study defines middlemen *conflicts with fishermen* as the degree of disagreements middlemen have in their transaction relationships with fishermen. Similarly, middlemen *conflicts with processors* is the degree of disagreements they have in their transaction relationships with processors.

Number of buyers: There is increasing attention about the contribution of network ties to the performance of economic transactions (e.g., Feld, 1981; Achrol, 1991; Heide, 1994;

Wathne and Heide, 2004). For example, networks ensure efficient transfer of information which enhances better adaptation to turbulent environments (Achrol and Kotler, 1999; Uzzi, 1997). Literature discusses network relations from different perspectives such as network density, centrality or number of ties (Antia and Frazier, 2001; Granovetter, 1973). This study considers the number of network ties which is defined as the number of buyers (processors) a middleman normally sells fish to. Although, it would be logical to consider number of suppliers, middlemen buy fish from a number of landing sites where fishermen sometimes operate in groups. This makes counting the number of suppliers difficult.

5.3 Hypotheses

Figure 5.1 hypothesizes that middlemen's preference for sustainability and quality- enhancing contracts will depend on the terms of contract in interaction with the context in which they operate.

5.3.1 The terms of contracts with fishermen and processors

Price information: Following TCE literature, middlemen tend to be well informed about markets and they benefit from information asymmetries between input and output markets (e.g., Ellis, 1988, Fafchamps, 2004). Middlemen may, on the one hand, deliberately create information asymmetries for their benefit and as such they may not be willing to give price information to fishermen to protect, for example, their bargaining advantage over price. On the other hand, however, access to price information is an important motivation for primary producers to engage in contracts to minimise price risks (Masakure and Henson, 2005). Hence, middlemen that may not give price information may lose suppliers which may endanger their business even more especially as fish supply is increasingly declining.

Therefore, although one would normally expect that middlemen would prefer more flexibility in terms of price adjustments to match market supply and demand, they may be compelled to give price information to secure supply. One way for middlemen to give price information is through fixed prices as is often the practice in farming (Grosh, 1994; FAO, 2001). However giving fixed prices may be counter productive for the middlemen if factory prices fall below prices they offer to fishermen. In order to avoid this scenario, middlemen may also need fixed prices from the processors. In that way, middlemen may figure out what prices to offer to fishermen. Hence, we envisage that middlemen would prefer to give and be given price information for them to engage in sustainability and quality-enhancing contracts. Following this thinking, we predict that:

H1: Middlemen will prefer sustainability- and quality-enhancing contracts with fishermen in which fish prices are fixed over contracts in which fish prices are fluctuating

H2: Middlemen will prefer sustainability- and quality-enhancing contracts with processors in which fish prices are fixed over contracts in which fish prices are fluctuating.

Selection of suppliers: Marketing literature demonstrates the strategic importance of and factors influencing selection of suppliers in competitive market environment. For example, Wathne, Biong and Heide, (2001) suggest that actors may switch to new partners if they offer better economic rewards than the current partners. For middlemen, selecting suppliers may be crucial because their performance in the output markets may depend on input supply. For example, if fishermen cannot guarantee fish quality, middlemen may experience more quality loss at the factories. Careful selection of contract fishermen may also minimise potential problems of adverse selection due to *ex ante* opportunism. For example, middlemen may select fishermen that may pretend to use sustainable fishing methods even when they are not. By careful selection of fishermen, middlemen may also minimise *ex post* opportunism whereby fishermen may undertake extra –contractual sell of fish. As Wuyts and Geyskens (2005) suggest, choosing close partners mitigates against opportunism. To that extent, middlemen may be better off contracting fishermen that already protect their interests than new ones whose behaviour cannot be ascertained. We thus predict that;

H3: Middlemen will prefer sustainability- and quality-enhancing contracts with fishermen they already know over contracts with new fishermen.

Period of contract: Contracts regardless of the period may lock middlemen into relationships with particular fishermen and/or processors with no chance to try alternative suppliers and/or buyers that may emerge within the period of contract. Hence the period of contract may have benefits and costs. For example, whereas short-term contracts may enable middlemen flexibility to easily switch fishermen and/or processors, they may incur high transaction costs such as renegotiating contracts, information and search costs (Williamson, 1985). In the purchase side, such transaction costs may not be substantial for middlemen to switch to new fishermen because of the high fishermen to middlemen ratio. In the supply side, it may be relatively difficult for middlemen to secure contracts with processors in the first place due to limited number of processors. Hence, they may have limited flexibility and incur high switching costs to switch from one processor to another. In that case, securing long-term contracts with processors may not only lower transaction costs for middlemen but most importantly, secure their businesses.

Under normal circumstances, middlemen would prefer more flexibility in deciding where to buy and sell fish to take advantage of competitive market conditions. More flexibility in the input market would not only enable middlemen to take advantage of competitive market conditions but also to switch from fishermen who do not follow sustainable and quality-enhancing practices as obliged. In the output markets, however, more flexibility would endanger middlemen's business. In view of the above scenarios, we envisage that:

H4: Middlemen will prefer short-term sustainability- and quality-enhancing contracts with fishermen over long-term contracts

H5: Middlemen will prefer long-term sustainability- and quality-enhancing contracts with processors over short-term contracts

Enforcement mechanisms: While enforcing compliance with the terms of contract is inevitable, compliance depends on whether enforcement mechanism exists to penalize breach of contract (Fafchamps, 1996). In view of the large number of small-scale middlemen in the Nile perch channel, there is a danger that middlemen that engage in sustainability- and quality-enhancing contracts may lose suppliers to their competitors that do not impose similar requirements. This may be true if processing factories, which rely on a number of small-scale middlemen for fish supply, do not enforce sustainable practices on all of them. As such, without proper mechanisms to enforce sustainable and quality-enhancing practices on all middlemen, they may not be motivated to engage in the contracts in the first place. One way to motivate middlemen would be for the processors not to buy fish from middlemen who do not promote sustainable practices. Alternatively, middlemen that do not promote sustainable practices should not be allowed to trade in the channel in the first place, i.e., they should lose their traders licenses. However, traders' license can only be revoked by government institutions that so far have not been effective in enforcing sustainable practices (see Chapter 3). Hence letting the same institutions punish middlemen that do not promote sustainable practices may not assure other middlemen that irresponsible middlemen are taken out of the channel. Hence, we envisage that middlemen would prefer that buyers, i.e., processors should only buy fish from middlemen that promote sustainable practices, i.e., refuse to buy fish that cannot be traced to fishermen when juveniles are detected. We therefore hypothesize that;

H6: Middlemen will prefer contracts in which processors only buy traceable fish over contracts in which middlemen may lose their licenses.

5.3.2 The moderating effect of the context

Supply uncertainty: The TCE contends that uncertainty is a key decision factor in choosing governance mechanisms (Williamson, 1985). Middlemen may face supply uncertainties when fishermen fail to catch adequate fish due to declining fish stocks, and further due to opportunistic behaviour when fishermen sell fish to other buyers. A middleman can manage supply uncertainty by using multiple suppliers concurrently and/or by switching suppliers until he finds reliable ones, i.e., without opportunistic intent (Joshi and Stump, 1999). However, as declining fish production signals increasing threats to sustainability, middlemen may prefer more flexibility in deciding where to buy not only to take advantage of competitive market conditions but also to ensure that they stop dealing with fishermen that show *ex post* opportunisms such as reverting to unsustainable fishing practices. In that way, middlemen may prefer relatively short-term contracts. Moreover, middlemen may switch from one fisherman to another with relative ease and low costs. So we envisage that:

H7: The middlemen that face high supply uncertainty, will prefer short-term over long-term sustainability and quality-enhancing contracts with fishermen

Quality losses: At the landing sites, quality of fish deteriorates as a result of lack of cooling facilities and poor fish handling such as throwing, beating or stepping on fish when packing or weighing. This means that middlemen who buy fish from fishermen who do not handle fish properly may incur more rejected fish at the factory. Although middlemen undertake sensory quality assessments, such techniques as colour of the eyes, gills, skin and firmness of the fillet may not reveal quality defects such as discoloration or crumbliness of the fillet that appear only after filleting when the fish may either be rejected or downgraded. Middlemen can minimise such quality losses by ensuring that fishermen handle fish properly or simply by engaging new fishermen who can guarantee quality. Since ensuring that fishermen handle fish properly may require time and effort to train and/or observe them, switching to new fishermen would be the probable response to higher quality losses. After-all, finding new fishermen may be relatively easy. So we expect that:

H8: Middlemen that incur more quality losses will prefer contracts with new fishermen over contracts with fishermen they currently transact with

Specific investment: Drawing from TCE, giving fishermen fishing gears and equipment may impose switching barriers for middlemen in order to protect their rents from expropriation by self interested fishermen (Rokkan, Heide and Wathne, 2003; Williamson, 1985). In that context, middlemen may prefer contracts with fishermen to whom they have given fishing gears to safeguard their investments. However, as Heide and John (1988) argue, contracts alone may not minimize opportunism under conditions of high uncertainties. In addition, literature demonstrates that contracts with small-scale primary producers often give both production facilities as well as market information (Masakure and Henson, 2005). Failure to give market information, especially price, may lead to extra-contractual sale of products especially when alternative market outlets exist that may offer better prices (Key and Runsten, 1999). In the Nile perch channel where competition is high due to declining production, extra-contractual sell of fish may be inevitable if fishermen are frustrated with for example, price risks. Hence, middlemen that have given fishing gears and equipment to fishermen would be in a better position to secure their rents through fish supply if they give price information as well. So, we predict that:

H9: Middlemen that give fishing gears to fishermen (a) will prefer sustainability- and quality-enhancing contracts with current fishermen and, (b) will prefer sustainability- and quality-enhancing contracts with fishermen in which fish prices are fixed

Trust: Marketing literature demonstrates that social relations such as trust minimize opportunism and enhances information sharing (e.g., Achrol and Gundlach, 1999; Joshi and Stump, 1999). This means that fishermen that act in mutual interest may be entrusted with sensitive information. Price information that is a critical source of bargaining advantage may not be easily shared without safeguards against opportunistic use of the information. In absence of any physical safeguards, trust offers an assurance that trustworthy partners seek mutuality and solidarity in the relationships (Cannon, Achrol and Gundlach, 2000). To that extent, we expect middlemen who strongly trust fishermen to provide price information. In addition, if they share price information to fishermen they trust, it seems logical that they will also prefer contracts with the same fishermen. Hence we hypothesize that:

H10: The more middlemen trust fishermen, (a) the more they will prefer sustainability- and quality-enhancing contracts with fishermen in which fish prices are fixed, and (b) the more they will prefer sustainability- and quality-enhancing contracts with current over new fishermen

Another important contribution of social relations is sustaining relationships. Although some literature suggest that trust substitutes contracts (Larson, 1992), other literature sources argue that social relations help to overcome the adaptive limits of contracts: a bilateral commitment to sustain the relations despite the unexpected complications and conflicts (Poppo and Zenger 2002). In the circumstances that middlemen operate, i.e., having limited alternative buyers, those that trust processors may thus be more motivated to seek long-term contracts. Therefore, we envisage that:

H 11: The more middlemen trust processors, the more they will prefer long-term over short-term sustainability- and quality-enhancing contracts with processors

Dependence: One aspect of dependence is that it may motivate partners to engage in successful and mutually beneficial exchange relationships by adhering to each others requests (Andaleeb, 1995; Morgan and Hunt, 1994). This means that middlemen that have higher dependence on fishermen are likely to adhere to their demands such as for price information to enhance mutual and reciprocal exchange relationships.

Another aspect of dependence is that a highly dependent actor may not accomplish important activities without the relationship (Argyres and Liebeskind, 1999; Hewett and Bearden 2001). For example, middlemen rely on processors for specialized tools such as refrigerated trucks for transporting fresh fish to the factory. Middlemen's reliance on processors is further enhanced by the (oligopsony) structure of the channel that leaves middlemen with limited alternative buyers. In such a scenario, securing a supply contract with a processor is for the middlemen like securing a livelihood because they would otherwise be out of business at least in the Nile perch channel. Hence, securing a long-term supply contract might be of utmost priority for middlemen to minimize business as well as livelihood uncertainties. With these views, we hypothesize that:

H12: (a) The more middlemen depend on fishermen, the more they will prefer sustainability- and quality-enhancing contracts with fishermen in which fish prices are fixed
(b) The more middlemen depend on processors, the more they will prefer long-term sustainability- and quality-enhancing contracts with processors

Conflict: How conflicts are managed when they arise is an important determinant for the continuation of the relationships (Hibbard, Kumar and Stern, 2001). The worst case scenario is when conflicts deteriorate into a spiral of hostility and distrust that ultimately could lead to dissolution of relationships (Ping, 1997). A middleman can manage conflicts with a fisherman either by switching to new ones (i.e. with whom they have no conflicts), or ignore conflicts and continue business as usual. While ignoring a problem may be an unlikely option, switching to new fishermen may depend on their availability which, for middlemen, may not be a problem. On the contrary, however since middlemen do not have many alternative buyers, switching may not be an option. Instead, seeking fixed prices in order to minimise price fluctuations over which conflicts occur might be a feasible option. In the context of the above scenarios, we envisage that:

H13: (a) Middlemen that tend to have a higher degree of conflicts with fishermen will prefer sustainability- and quality-enhancing contracts with new over current fishermen
(b) The higher the degree of conflicts middlemen have with processors, the more they will prefer sustainability- and quality-enhancing contracts with processors in which fish prices are fixed

Number of buyers: Building on literature on network ties (e.g. Granovetter, 1973), a middleman who has many buyers might be in an advantage position in terms of accessing price information. By implication such middlemen may not necessarily have to develop contractual relations with particular buyers to access price information. In addition, middlemen that have many buyers spread risks of market uncertainties. As such, they may not want to be tied up to a particular processor for long in the event that competitive conditions emerge from other processors. For example, as fish supply declines, processors may offer competitive conditions to attract suppliers. Hence, middleman with many buyers may prefer short-term contracts with processors so that they may easily switch to emerging competitive buyers. Hence, we predict that

H14: (a) Middlemen that supply fish to many processors will prefer contracts with processors in which fish prices are fluctuating.
(b) The more buyers middlemen supply fish to, the more they will prefer short-term (6 months) contracts with processors

The hypotheses are summarised in Table 5.1.

5.4 Methodology

In order to test our hypotheses, data was collected from middlemen that buy fish from fishermen and supply to processing factories. In this section, we describe data collection procedures, measurements and validation.

Table 5.1. Summary of hypotheses.

Factor	Factor level	Main effect	Moderating effects						
			Supply uncertainty	Quality losses	Fishermen given gears	Trust	Dependence	Conflict	Number of buyers
Price information for fishermen	Fixed prices	(+) H1			(+) H9b	(+) H10a	(+) H12a		
	Fluctuating prices								
Factory prices	Fixed prices	(+) H2						(+) H13b	
	Fluctuating prices								(+) H14a
Selection of supplier	Current fishermen	(+) H3			(+) H9a	(+) H10b			
	New fishermen			(+) H8				(+) H13a	
Period of contract with fishermen	6 months	(+) H4	(+) H7						
	1 year								
Period of contract with processor	6 months	(+) H5							(+) H14b
	1 year								
Enforcement mechanism	No sell of fish without traceability	(+) H6				(+) H11	(+) H12b		

5.4.1 Procedure

Sampling: In order to identify our sample, we approached the fish processing factories in Kenya to obtain a list of the middlemen that supply fish to them. However, only one factory was cooperative to give a list of its suppliers while others did not cooperate. No specific reason was given why some processors were not willing to give us a list of their suppliers. But probably, they thought that we might be asking the middlemen on sensitive issues about the factories. As a result, other middlemen were identified at the beaches where they were buying fish; on their way to/from the beaches and also in convenient places such restaurants or rest houses. The problems in identifying and tracing respondents are typical in developing countries (see Poate and Daplyn, 1993).

The survey: Data was collected through a questionnaire which consisted of multi-item scales measuring trust, dependence, conflict, supply uncertainty; quality losses, demographic characteristics and a conjoint analysis task. The interviews were conducted in English and two local languages, i.e., Luo, the language of one of the local tribes and Swahili, the trade language for East Africa). The questionnaires were pre-tested once with 5 middlemen before the final data collection. No major problems were noted during pre-testing. All the item measurements were evaluated on a 5 point Likert scale scales where the score of 1 meant strongly disagree and 5 was strongly agree.

The sample: Out of the middlemen approached five did not participate in the survey. Three middlemen indicated to have no time for the interviews as we found them when they were leaving the beaches to deliver fish to the factories. The other two middlemen were leaving the factories for the beaches. Finally a total of 48 middlemen participated in the interviews. In order to minimise selection bias (Blair and Zinkhan, 2006), any middlemen we could identify was approached for an interview. Due to their mobility, it was however difficult to interview many middlemen. However, from our informal enquiries from those interviewed, there could be about 60-70 middlemen supplying fish to the (six) processors at the time of the study. The middlemen were of ages ranging from 20 to 52 years with 33 middlemen (69%) between 30 to 40 years while the remaining 15 (31%) middlemen were below 30 and above 40 years. Twelve (25%) had some years of primary education, 30 middlemen (i.e. 62.5%) had some years of secondary education, and 6 middlemen (12.5%) had some tertiary education. Forty four middlemen (about 92%) were only buying and selling Nile perch while 4 (8%) were selling other fish types also. Twenty nine (about 61%) had other major income generating activities besides fish trading while the rest (19) did not have other major income generating activities apart from fish trading. Out of the 48 middlemen, 45 (94%) had loaned fishing gears to fishermen. Ten middlemen (i.e. 20%) had kinship relations with fishermen from whom they were buying fish (e.g. brother, sister, cousin) at the time of the study.

Conjoint analysis task: In order to determine how middlemen could develop preferences for hypothetical contracts, a conjoint analysis was used (Green and Srinivasan, 1990). The hypothetical contract attributes were developed based on theoretical framework (Figure

5.1). In order to implement the conjoint analysis, a full profile presentation method often recommended for up to 9-10 factors depending on the complexity of the attributes (Green and Srinivasan, 1978, 1990, Hair et al., 1998) was used. With a 2^6 full factorial design, middlemen would have to evaluate a total of 64, i.e., 2*2*2*2*2*2 hypothetical contracts excluding holdout profiles. In order to minimise information load and boredom, a fractional factorial main effects design was used to reduce the number of profiles to 8. In total, middlemen evaluated 12 contracts which included four holdout tasks (Hair et al., 1998).

Setting the context for the conjoint tasks: In order to implement the conjoint analysis task, personal interviews were used. Middlemen were briefed about the purpose of the task, what the attribute levels meant and how profiles were to be evaluated. In the introduction, middlemen were also asked to consider the real life situation, i.e., the degradation of the Nile perch and the quality of the fish they buy and sell. Then, they were asked to imagine that they were being approached to sign a contract that oblige them to promote sustainability and quality by only buying and supplying fresh fish of not less than 1.5kgs according to the Fisheries Act of Kenya (Kenya Government, 1991). Then they were presented with the conjoint profiles one at a time for which they were asked to indicate the extent to which they would be willing to sign up each contract. Profiles were to be rated on a Likert scale of 1 to 5, where 1 was least willing and 5 was most willing (see Box 5.1 for an example of a profile).

5.4.2 Measurements and validation
In order to measure the constructs for trust, conflict and dependency, we adapted existing multi-item scales (e.g. Brashear et al., 2003; Andaleeb, 1995; Jap and Ganesan, 2000) (see appendix 5.1). The measurement items for supply uncertainty were new. Quality losses measured the proportion of fish rejected at the factories. Specific investment measured the investment that

---

Box 5.1. An example of a conjoint profile.

Profile
Factory prices will be changing anytime without notice
This is a 6 months contract with a processor
Middlemen that do not identify fishermen when they buy and sell undersize fish will not sell their fish
You will sign this contract with your current fishermen
You will change prices for fishermen anytime without telling them
This is a 6 months contract with the fisherman

Question: To what extent would you be willing to sign this contract for you to only buy and supply fresh Nile perch fish of at least 1.5kgs?

Least willing   1   2   3   4   5   Most willing

---

middlemen have done to secure fish supply. The number of fishermen given fishing gears and equipment was used as a proxy for the specific investments. In order to validate the multi-item scales for trust, conflict, dependence and supply uncertainty, recommendations given by Shook *et al.* (2004) were followed i.e., to examine unidimensionality, discriminant validity and reliability.

First, the measurements were examined for unidimensionality through exploratory factor analysis (Churchill, 1979). All items that loaded on multiple factors and/or with low loadings were dropped after varimax rotation (Hair *et al.* 1998). In order to assess if the retained items were significantly contributing to constructs being measured, they were imputed into a confirmatory factor analysis model in LISREL 8.72 (Jöreskog and Sörbom, 2005) with all factors. The factor loadings were examined and were all significant (t>1.96) (Byrne, 1998). For all the models, the Comparative fit index (CFI) was above .95, the Goodness of Fit Index (GFI) and Adjusted Goodness of Fit Index were above .90; and Root Mean Square Error of Approximation was ≤ .05. These statistics suggest good model fit (Schermelleh-Engel and Moosbrugger, 2003).

After establishing the unidimensionality, discriminant validity testing was undertaken. To do that, constructs were examined in a series of two - factor confirmatory models (Anderson, 1987; Bagozzi and Phillips, 1982). Each model was run twice, first, constraining the covariance and variances to 1 and then removing the constraint. Following this, a Chi-square difference and changes in the Comparative Fit Index (CFI) (Byrne, 1998) were examined. For all models investigated, the Chi-square values were significantly lower for the unconstrained models than for the constrained models and the CFI values for the constrained models were lower suggesting poor fit. The changes in Chi-Square and CFI are given in Appendix 5.2. Finally, in order to verify the discriminant validity results, the constructs were further examined following the procedure given by Fornell and Larcker, (1981). This procedure suggests that the average variance extracted from each item by the construct should be greater than the shared variance between the constructs. This was confirmed in all our models (see Appendix 5.3). The reliability of the constructs as determined by the Cronbach alpha was all good. Table 5.3 and 5.4 give a summary of the variables used in this study.

5.4.3 The conjoint model
In order to validate the conjoint tasks, the data was examined for any missing information. There were no peculiarities or missing information on the profiles. Then we performed a conjoint analysis. To assess the predictive validity of the model, first, we assessed how the model fit the data and, we further assessed the predictive validity in predicting the holdout sample. The model fit was assessed by examining the Pearson correlation coefficients that give the correlation between the original and predicted preference scores. The Pearson correlation coefficients (Figure 5.2) show that the model adequately represented the data with only one questionnaire that had low Pearson correlation coefficient. That questionnaire was dropped leaving 47 respondents that had an average Pearson coefficient of .8103 with a standard deviation of .086.

Table 5.3. Descriptive statistics of the purified constructs and other variables.

Variables	Operationalization	No. of items	Range	Mean	Std. Dev.	Cronbach alpha
1 Trust in fishermen	Middlemen's belief that a fisherman is sincere, fulfils promises and of reputation (Geyskens, Steenkamp and Kumar, 1998)	4	1-5	4.22	0.79	0.88
2 Trust in processors	Middlemen's belief that a processors is sincere, fulfils promises and of good reputation (Geyskens, Steenkamp and Kumar, 1998)	4	1-5	3.85	1.08	0.91
3 Dependence on fishermen	Extent to which middlemen rely on fishermen in their business (Hewett and Bearden, 2001)	3	1-5	2.63	1.04	0.83
4 Dependence on processors	Extent to which middlemen rely on processors in their business	2	1-5	2.65	1.21	0.81
5 Conflict with fishermen	Degree of disagreements with main supplier in matters relating to transaction relationships	2	1-5	3.43	0.62	0.68
6 Conflict with processors	Degree of disagreements with processors on issues relating to transaction relationships	2	1-5	2.80	1.01	0.79
7 Supply uncertainty	The extent to which middlemen may predict daily fish supply relative to five years ago	2	1-5	4.36	0.83	0.85
8 Quality losses	Proportion of fish rejected at the factory (%)		1-35	6.42	6.85	
9 Number of fishermen given gears	Number of fishermen given fishing gears and boats by middlemen		3-60	15.64	15.04	
10 Number of buyers	Number of processors middleman normally sells fish to		1-3	1.17	.519	

Table 5.4. Correlation matrix.

Variables	1	2	3	4	5	6	7	8	9
1 Trust in fishermen									
2 Trust in processors	.164								
3 Dependence on fishermen	-.230	-.200							
4 Dependence on processors	-.099	.064	.187						
5 Conflict with fishermen	.034	-.263	-.176	-.140					
6 Conflict with processors	-.247	-.410**	.108	.087	-.248				
7 Supply uncertainty	-.219	-.238	.102	.199	-.030	-.117			
8 Quality losses	-.136	-.139	.017	-.042	-.244	.267*	-.270*		
9 Number of fishermen given gears	.101	.295*	-.098	.144	.093	-.125	.076	-.173	
10 Number of buyers	.169	-.025	-.166	-.088	.182	-.099	-.084	.055	.183

* Significant (p< .05) ** Significant (p<.01) (2-tailed).

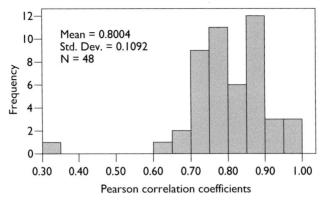

Figure 5.2. Pearson correlation coefficient for assessing model fit for original sample.

Next, the remaining 47 questionnaires were examined for predictive validity by assessing the ability of the model to predict the holdout sample (Green and Srinivasan, 1990). To do that, we examined the Kendall's tau for the holdout sample. The Kendall's tau revealed that for 4 respondents, the model did not generalise beyond the sample, i.e., they had low predictive validity (see Figure 5.3). Using a cut-off of .30 as reasonable, these respondents were also dropped from the sample leaving a final sample of 43 middlemen that had an average Kendall's tau of .702 with a standard deviation of .144. The low predictive validity for the few middlemen could be attributed to the possible inconsistencies in translation of the conjoint profiles or that the respondents might not have been serious in evaluating the profiles.

5.4.4 Data analysis
After respondents were individually scrutinised in order to remove those that were not reliable, we carried out an aggregate analysis to test our hypotheses. The ordinary least squares regression

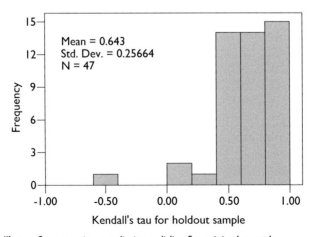

Figure 5.3. Kendall's tau for assessing predictive validity for original sample.

analysis was used to test the hypotheses. The analysis was done in sequence; (1) determining the types of contracts they prefer and, (2) the moderating effects on their preference for particular contracts.

Determining the type of contracts middlemen prefer: In order to determine the types of contracts middlemen prefer, a regression analysis was run to explain the preference for contracts as a function of the terms of contracts, i.e., main effects. In this regression, preference scores for the eight calibration profiles were standardized to remove the possible effects of any extreme preferences for particular profiles. Standardisation of the profiles makes the preference scores for the calibration profiles comparable across middlemen. Since the preference for each profile is the unit of analysis, it implies that the overall sample was 43 * 8 = 344 where 43 was the number of valid respondents in the final sample and 8 was the number of calibration profiles. To run the actual regression, the attribute levels were coded following an effects coding scheme (Cohen and Cohen, 1983) in order to represent the different levels of the factors in the regression analysis. In this scheme, the first level of each factor (e.g., fixed prices for fishermen) was coded as −1, and the other (e.g., fluctuating prices for fishermen) as +1. These attribute level dummies were then used as explanatory variables for the standardized preference scores. Model fit statistics were examined.

Determining moderation effects: In order to determine whether middlemen's preferences for particular contracts were moderated by the contextual variables both in their demand and supply markets, a second regression analysis was run. The moderating variables were defined by multiplicative products between the attribute level dummies and the respective contextual factors. Again, the standardized preference scores for particular contracts were estimated as a function of the terms of contract (main effects) and their interactions with contextual variables. The significance of the change in R^2 as a result of the inclusion of the moderating effects was tested. This was necessary to ascertain if the change in the variation in preference for particular contracts explained by the inclusion of moderating effects was significant.

Before the actual analyses, the independent variables were mean centred to remove the possible effects of multicollinearity (Aiken and West, 1991). The variance inflation factor (VIF) and tolerance for the regression models were nonetheless examined in both regression models (Hair *et al.*, 2001). The VIF were below 2 i.e., below the cut-off point of 10 suggested in the literature (Hair *et al.*, 2001). In addition, we tested for heteroskedasticity i.e., the problem of unequal variance (Pindyck and Rubinfeld, 1998). To do that, we tested the relationship between predicted preferences and standardized residuals in both models (Maddala, 1992) which showed no relationship.

5.5 Results

This section presents results in the order of the analysis. First, the types of contracts that middlemen prefer are given and second, how middlemen's preferences for particular contracts were moderated by the context in which middlemen operate.

5.5.1 The type of contracts middlemen prefer

In order to determine the types of contracts that middlemen prefer we estimated the main effects of the terms of contracts on preference for contracts. We also examined the utility distributions. The results are shown in Table 5.5.

The regression model explains about 19% of the variation in the middlemen's preference for contracts. This variation is mainly explained by the fixed price for fishermen, short term (6 months) contract with fishermen, long term (1 year) contracts with processors and contract with current fishermen. The results show that middlemen prefer contracts with fishermen in which prices are fixed - consistent with Hypothesis 1 (p<.01). An examination of the individual level utilities (see Appendix 5.4) was done. Using a cut off point of -1 and +.1, the utility distribution shows that 32 (74%) middlemen preferred contracts with fishermen in which fish prices are fixed compared to 10 (23%) middlemen who prefered fluctuating prices and 2 middlemen did not care.

Table 5.5. Regression coefficients for the main effects of terms of contract (N=343 contracts).

Independent variables (factor level dummies)	Dependent variable: Standardized preference scores		
	Unstandardized coefficients	Hypothesis	Remarks
Terms of contracts with fishermen			
Fixed price for fishermen	.259***	(+) H1	Supported
Contract with current fishermen	.059 *	(+) H3	Supported
6 months contract with fishermen	.292 ***	(+) H4	Supported
Terms of contracts with processors			
Fixed factory prices	.031	(+) H2	Not supported
1 year contract with processor	.098**	(+) H5	Supported
No sale of fish without traceability	.023	(+) H6	Not Supported
Statistics	R^2 (Adj. R^2)	.191 (.177)	
	F	11.58***	

*** Significant (p<.01) ** Significant (p<.05) * Significant (p<.10) (one - tailed)

We expected in Hypothesis 2 that middlemen would prefer contracts with processors in which fish prices are fixed. However, the results do not show any support for our expectation (p>.10). In order to explore the results further, the utility distribution for fixed factory prices show that middlemen were almost split over fixed or fluctuating factory prices. Seventeen middlemen (about 39%) preferred contracts with processors in which fish prices were fixed compared to 18 middlemen (about 42%) that prefer contracts with processors in which fish prices were fluctuating. The remaining 8 middlemen (about 19%) were indifferent between fluctuating or fixed prices.

In Hypothesis 3, we expected that middlemen would prefer contracts with fishermen they already know, i.e., current fishermen. The results give modest support for the Hypothesis (p<.10). An exploration of the utilities shows that about 29 (67%) middlemen prefer contracts with current fishermen. However, the low significance could be a result of about 7 middlemen (16%) that had very high preference for new fishermen. Six middlemen (14%) did not care about whether they contract new or current fishermen.

The results also show that middlemen prefer 6 months contracts with fishermen. These results are consistent with Hypothesis 4. The utility distribution for the 6 months contract with fishermen also demonstrate that about 24 middlemen (56%) prefer 6 months contracts with fishermen while about 14 (33%) middlemen were indifferent about the period of contracts with fishermen. Six middlemen (about 14%) prefer 1 year contracts with fishermen.

Hypothesis 5 predicted that middlemen would prefer long-term contracts with processors. The results show that middlemen prefer one year contracts with processors which supports the prediction (p<.01). The utilities show that 20 middlemen (47%) respondents prefer 1 year contracts with processors compared to 10 (23%) middlemen that prefer 6 months contracts with processors. About 13 middlemen (30%) were indifferent about the period of contract with processors.

Hypothesis 6 predicted that middlemen will prefer contracts in which middlemen that do not trace their fish would not be allowed to sell their fish. The results do not support the prediction (p>.10). This could be explained through an examination of the distribution of the utilities that show that the majority i.e., 28 middlemen (65%) actually did not care about enforcement mechanisms. It was only about 12 middlemen that prefer contracts in which middlemen that do not trace their fish back to fishermen should not sell fish. The remaining 3 middlemen prefer contracts in which middlemen that do not trace their fish should have their licenses withdrawn.

In short, the results show that there are differences in the middlemen preferences for the types of the contracts. We assume that such differences might be explained by the moderating factors that reflect the context in which middlemen operate. We test this in the next section.

5.5.2 The moderating effects of the contextual variables

The main effects show that middlemen prefer different contracts such as short-term contracts with fishermen, long-term contracts with processors and contracts with fishermen in which fish prices are fixed. We assume that such variation in preferences could be influenced by the circumstances surrounding middlemen. In order to test this assumption, a regression analysis was run in which preference for contracts was explained as a function of the terms of contract and the moderating variables. The model explains about 33% of the variation in middlemen's preference for contracts. The difference in the R^2 between the main effects and interaction effects models is significant {$F_{(13, 249)} = 3.25; p<.01$}. Table 5.6 gives the results.

The results show that the more middlemen face supply uncertainty, the more they prefer short-term (6 months) contracts with fishermen. These results support ($p<.05$) Hypothesis 7. Although as a main effect, we expected that middlemen would generally prefer contracts with fishermen they already know, i.e., current fishermen, the main effect is not significant ($p>.10$). However, the moderating effects show that the more middlemen give fishing gears to fishermen, the more they prefer contracts with current fishermen, which is consistent with Hypothesis 9a. Furthermore, the results also show that the more middlemen trust fishermen, the more they would prefer contracts ($p<.05$) with fishermen they already know - consistent with Hypothesis 10b. The results show that middlemen who incur more quality losses when they deliver fish to the factory will instead prefer contracts with new fishermen ($p<.05$). This is consistent with Hypothesis 8.

In complement to the preceding results, the results show that middlemen that give fishing gears to fishermen prefer contracts with fishermen in which fish prices are fixed ($p<.01$). This is in line with Hypothesis 9b. Furthermore, the results show that the more middlemen trust fishermen, the more they would also prefer contracts with fishermen in which fish prices are fixed ($p<.01$). This is consistent with Hypothesis 10a. Also in support of Hypothesis 12a, the results show that the more dependent middlemen are on fishermen, the more they will prefer contracts with fishermen in which fish prices are fixed ($p<.05$).

In Hypothesis 11, we assumed that the more middlemen trust processors, the more they would prefer long-term (1 year) contracts with processors. We also assumed in H12b that the more dependent middlemen are on processors, the more they would prefer long-term contracts with processors. However there is insufficient evidence beyond the main effect to suggest that trust in and dependence on processors enhance middlemen's preference for long-term contracts with processors ($p>.10$). Hence Hypotheses 11 and 12b have not been supported.

The result further show that middlemen that tend to have a higher degree of conflicts with fishermen would prefer contracts with new fishermen ($p<.05$). These results support Hypothesis 13a. We also assumed in H13b that the higher the degree of conflicts middlemen have with processors, the more they will prefer contracts with processors in which fish prices are fixed. However, the results do not give sufficient evidence to support this assumption ($p>.10$).

Table 5.6. Regression coefficients for moderating effects for preference for contracts (N=311 contracts).

Independent variables	Unstandardized coefficients	Hypothesis	Remarks
Terms of contracts with fishermen			
Fixed prices for fishermen	.248***	(+) H1	Supported
Contract with current fishermen	.055	(+) H3	Not supported
6 months contract with fishermen	.324***	(+) H4	Supported
Terms of contracts with processors			
Fixed factory prices	.036	(+) H2	Not supported
1 year contract with processors	.092**	(+) H5	Supported
No sale of fish without traceability	.020	(+) H6	Not Supported
Moderating effects in the purchase side			
6 months contract with fishermen* Supply uncertainty	.187**	(+) H7	Supported
Contract with new fishermen * Quality losses	.262**	(+) H8	Supported
Contract with current fishermen * Fishermen given gears	.048**	(+) H9a	Supported
Fixed prices for fishermen * Fishermen given fishing gears	.133***	(+) H9b	Supported
Fixed prices for fishermen *Trust in fishermen	.153***	(+) H10a	Supported
Contract with current fishermen *Trust in fishermen	.106 **	(+) H10b	Supported
Fixed prices for fishermen * Dependence on fishermen	.092**	(+) H2a	Supported
Contracts with new fishermen * Conflict with current fishermen	.108**	(+) H13a	Supported
Moderating effects in the supply side			
1 year contract with processor *Trust in processors	.069	(+) H11	Not supported
1 years contract with processors * Dependence on processors	.068	(+) H12b	Not supported
Fixed factory prices * Conflict with processors	.003	(+) H13b	Not supported
Fluctuating factory prices * Number of buyers	.092*	(+) H14a	Supported
6 months contract with processors * Number of buyers	.171**	(+) H14b	Supported
Statistics	R^2 (Adj R^2)	.325 (.281)	
	F	10.00***	

*** Significant (p<.01); ** significant (p<.05) *significant (p<.10) (one - tailed)

Hypothesis 14a predicted that middlemen that supply fish to many buyers would prefer contracts with processors in which fish prices are fluctuating. These results marginally (p<.10) support the hypothesis. Although the results marginally support the hypothesis, they could partially explain the insignificant main effect for the fixed factory prices in view of the fact that middlemen were equally divided. It could also be that middlemen already have some form of contracts with processors through which they obtain price information. Consequently, it might not have been a major motivating factor beyond the way they already obtain price information. Finally, the results show that the more buyers the middlemen supply fish to, the more they will prefer short-term (6months) contracts with processors. This is in line with Hypothesis 14b.

5.6 Discussion and implications

5.6.1 Middlemen preference for contracts

The objective of this chapter was to investigate factors that may influence middlemen's preference for contracts that oblige them to promote sustainable and quality-enhancing practices in their buying and supplying activities.

In the buying side, the results show that middlemen in general and in particular, those that *invest* in fishermen, *trust* fishermen, and are more *dependent* on fishermen prefer contracts in which fish prices are fixed. These results can be understood from the fact that middlemen may, first and foremost, want to secure fish supply which is increasingly uncertain as fish production continues to decline. In fact, middlemen invest in fishermen in order to secure fish supply (Chapter 3). Hence middlemen may want to retain (or attract) suppliers by giving them price information. Middlemen may further want to protect their investments including the information that they may give from fishermen's opportunism. That is why they prefer to give price information to trustworthy fishermen. Middlemen that are more dependent on fishermen may provide price information in order to secure the relationship they depend on. After all, according to Hewett and Bearden (2001), highly dependent middlemen may not fulfill their business activities without the relationship with particular fishermen. Hence, it is logical that such middlemen need to secure their relationships by, in this case, providing price information.

The results show that the effect of selecting contract suppliers, i.e., between *current* and *new* fishermen, was indicatively significant in the main effects only model. The effect is not significant in the moderating effects model. Interestingly however, middlemen are split over contracting current or new fishermen. On the one hand, middlemen that incur more *quality losses* and also those that tend to have a higher *degree of conflicts* prefer contracts with *new fishermen*. This is understandable because without good quality, middlemen may not be competitive in their transactions with processors. Similarly, middlemen would naturally avoid engaging into contracts with fishermen they obviously have conflicts with. On the other hand, middlemen that trust fishermen and have also invested in fishermen prefer to have contracts with the same fishermen. This means that contracts may not necessarily disrupt existing relations with fishermen, instead they may strengthen some of them.

Further, the results show that middlemen generally prefer *short-term* contracts with fishermen. This is particularly the case for middlemen that experience increasing supply uncertainty. Such middlemen may want to switch suppliers in the event that the contracted fishermen are opportunistic both in terms of having extra-contractual sell of fish or reverting to unsustainable fishing practices. This means that fishermen that do not comply with the terms of contracts, i.e., sustainable and quality - enhancing practices may easily be identified and left out of the channel. Such fishermen may later be compelled to engage in sustainable practices or remain out of business, i.e., out of the international channel.

In the supply side, the results show that middlemen generally prefer *long-term* contracts with processors. This can be explained by the fact that it may be difficult for the middlemen to secure supply contracts due to the limited number of buyers. Hence, middlemen may want to secure long-term contracts to minimize risks of failing to get another supply contract. This is in fact supported by the fact the middlemen that currently secure multiple contracts prefer short-term contracts with processors. For these middlemen, the risks of having no supply contracts may be small. Further such middlemen may not want to be locked up to a particular processor for long in the event that they may miss emerging competitive conditions with other buyers.

Although the main effects model does not show significant preference for fixed factory price, the moderating effect model shows that middlemen that supply fish to *more than one buyer* prefer contracts with processors in which fish prices are fluctuating. This is understandable from the network theory (e.g., Granovetter, 1973) that such middlemen may get price information anyway from other network nodes. Such middlemen may want to take advantage of the price differences between buyers in order to decide to supply more fish and when.

Unlike in the buying side, in the supply side, middlemen's social relations, i.e., *trust, dependence* and *conflict* with processors do not significantly moderate middlemen's preference for the terms of contracts. We can attribute these results to the structure of the chain. According to our estimates, about 60-70 middlemen supply fish to 6 processing factories in Kenya. Although each processing factory relies on a number of middlemen, this structure alone gives much power to the processing factories that may disadvantage middlemen. For example, processors may easily engage or disengage middlemen because as Heide and John (1992) note, processors may use their structural power (i.e., market concentration) to exercise control over middlemen. In return, whether middlemen have good or bad relations with processors might not matter as long as they (middlemen) secure their businesses and more so, their livelihoods. With such a structural power for processing factories, their social relations with middlemen may in fact help them to select the middlemen that suit their interests. The implication of these results could be that middlemen especially those that cannot secure multiple contracts may be desperate to accept any contract terms the processors may offer provided it secures their livelihoods. Alternatively, it could be that middlemen might think that whether or not they prefer better terms of contracts, processors will still offer what they want anyway.

5.6.2 Middlemen/fishermen relations

In view of the similarities of some of the contract attributes and moderating effects tested for fishermen and middlemen, the results of this chapter raise some implications for the feasibility of contracts to promote sustainability and quality-enhancing practices between middlemen and fishermen. Table 5.7 summarises some of the common factors between fishermen and middlemen results.

One of the motivating factors for fishermen to engage in sustainability and quality-enhancing contracts is access to price information. The fact that middlemen also generally prefer contracts with fishermen in which fish prices are fixed implies that middlemen and fishermen may, on that note, engage in sustainability and quality- enhancing contracts. In addition, conflicts between fishermen and middlemen increased fishermen's preference for contracts in which fish

Table 5.7. Preferences for common terms of contracts for fishermen and middlemen.

Terms of contract	Fishermen preference	Moderating factors	Middlemen preference	Moderating factors	Remarks
Price information	Fixed prices	Conflict with middlemen	• Fixed prices with fishermen • Fluctuating factory price	• Conflict with fishermen • Dependence on fishermen • Number of buyers	• Middlemen and fishermen may engage in contracts • Conflicts between middlemen and fishermen may be minimized
Selection of contract partners	Processors	Conflict with middlemen	• New fishermen • Current fishermen	• Conflict with current fishermen • Quality loss • Investment in fisherman • Trust in fishermen	• Some middlemen and fishermen may engage in contrac • Some fishermen may engage in contracts with processors
Enforcement	Buyers should not buy fish caught with bad gears		• Results not significant		• Fishermen prefer private policy enforcement sustainable fishing practices

prices are fixed and also contributed to fishermen's preference for contracts with processors. The fact that middlemen also prefer contracts with fishermen in which fish prices are fixed implies that conflicts between fishermen and middlemen (often over fish prices) would be minimised. In that way, fishermen and middlemen may understand each other and promote sustainable and quality-enhancing practices for their mutual benefits.

However, conflict of interest may come in with fishermen's strong preference for contracts with processors unlike middlemen. Fishermen prefer contracts with processors irrespective of whether they trust middlemen. Fishermen's strong preference for contracts with processing factories could be a result of expected economic benefits. Fishermen's intuition as well as knowledge that middlemen, after-all, benefit from the processors may influence their expectation for better gains with direct transactions with processors. Hence, although fishermen have for long depended on middlemen for fish markets and to some extent, access to fishing gears, they may just need to try alternative market outlets now that a chance arose with contracts. This is similar with middlemen who incur more quality losses and have high conflict with the fishermen they currently transact with prefer contract with new fishermen.

In short, whatever the motive behind the fishermen's preference for contracts with processors, this conflict of interest in selection of contract partners may endanger middlemen's position in the channel. The implication of these results would be that some fishermen may engage in contracts with middlemen while others may engage in contracts with processors, i.e., assuming that processing factories are willing to do so.

5.7 Study limitations and future research

Despite the support for our conceptual framework, a number of issues may be raised for improvements in future research. The study relied on the theoretical framework of TCE, social and networks on the presumption that information asymmetry and opportunism that often motivate the use of contracts can be addressed. However, these arguments have not been adequately supported especially in the supply side of the middlemen. For example, although the TCE argues that middlemen would engage in contracts to minimise information asymmetries (Williamson, 1985), we find support for that in the purchase side of middlemen but not in the supply side. In the supply side, middlemen not only face an oligopsony market but also buyers of much more larger scale of operation. Hence future research may thus consider other contextual factors as well as contract attributes to investigate what may motivate actors in an oligopsony facing large-scale actors to engage in contracts.

Although arguments for the relationships between social relations such as trust and contracts have been conflicting (see Larson, 1992; Uzzi, 1997), we followed the argument that trust and contracts could be complementary (Poppo and Zenger, 2002) such that trust would positively moderate preference for particular contracts with fishermen and processors. Whereas we find strong moderating effect for trust in the purchase side, we do not find any moderating effect in the supply side of the middlemen. Similarly, we find no moderating effect for dependence and conflicts in the supply side. Future research may consider more variables such as cooperation, satisfaction and power relations to establish what might influence the type of contracts actors that face an oligopsony and comparably very large-scale transaction partners might prefer.

Another point that this study could not establish is the interactions within the model. Although, it was logical that interaction within the model could have been tested, i.e., whether the preferences for particular contracts in the supply side could have influenced or been influenced by preferences in the purchase side, the fractional factorial design that was used in the study analyzes main effects only. Allowing for interactions could have increased the number of contracts the middlemen could have evaluated. We therefore consider this study as an exploratory study whose findings may be enhanced by more in-depth future research. Future research should therefore consider possible interactions which could perhaps give more insights in the effect of the double loyalty that middlemen face. However doing so would not only increase the number of profiles but it would also need different conjoint approaches such as the adaptive or the trade-off methods that are capable to handle more profiles (Hair *et al.*, 2001).

Another limitation is the sample size. While it was not possible to get more than 48 middlemen, the fact remains that sample size is relatively small and it may have consequences on the generalizability of the results. Although we informally estimated that there could be about 60-70 middlemen operating in the channel in Kenya, no one knows for sure, the exact number of middlemen in the chain. It is therefore important that further studies put more time than rendered in this study to trace most, if not all, middlemen.

Appendices

Appendix 5.1. Multi-item measurements scales

This appendix provides information on the measurement scales i.e. source of the scale, factor loadings, T-values, eigenvalues for items and Cronbach alpha.

1. Trust on fishermen (alpha = 0.875; eigenvalue = 2.902)	Factor loadings	T values
My fisher is very sincere	0.86	6.66
My fisher has a good reputation	0.78	5.68
My fisher always fulfils his promise	0.87	6.76
My relationship with this fisher is satisfactory	0.67	4.69
All items adapted from Andaleeb, (1995)		

2. Trust on processors (alpha = 0.907; eigenvalue = 3.147)	Factor loadings	T values
This processor is very sincere	0.91	7.74
This processor has good reputation	0.87	7.26
This processor always fulfils his promise	0.75	5.74
My relationship with this processor is satisfactory	0.85	6.93
This processors is very dependable	Dropped	
All items adapted from Andaleeb, (1995)		

3. Dependence on fishermen (alpha = 0.825; eigenvalue = 2.25)	Factor loadings	T values
I can easily find other fishers to buy fish from (R)	0.80	5.88
I fully depend on this fisher for my business	0.68	4.85
I can easily buy fish even if this fisher stops selling fish to me (R)	0.90	6.75
I can easily find a better fisher than the current one (R)	Dropped	
All items adapted from Jap and Ganesan, (2000)		

4. Dependence on processors (Alpha = 0.805; eigenvalue 1.674)	Factor loadings	T values
I can easily find another factory to buy my fish (R)	0.55	2.08
I can easily sell fish even if this factory stops buying from me (R)	0.93	2.34
I fully depend on this factory	Dropped	
I can find a better factory than this current one	Dropped	
Items adapted from Jap and Ganesan, (2000)		

5. Conflict with fishermen (Alpha = 0.675; eigenvalue = 2.48)	Factor loadings	T values
The relationship between me and this fishermen is very good (R)	0.76	8.91
We have significant disagreements in our business relationship	0.73	7.44
We frequently quarrel with this fishermen over issues of our business	Dropped	

Adapted from Jap and Ganesan, (2000)

6. Conflict with processors (Alpha = 0.785; eigenvalue = 2.98)	Factor loadings	T values
The relationship between me and this processor is very good (R)	0.67	10.95
We have significant disagreements in our business relationship	0.73	9.24
We frequently quarrel with this processor issues of our business	Dropped	

Adapted from Jap and Ganesan, (2000)

4. Supply uncertainty [c] (alpha = 0.85; eigenvalue = 1.744)	Factor loadings	T values
It is easily to know how much fish supply one would get per trip now than it was five years ago (R)	0.76	3.24
It is always certain how long one may take to load a truck than five years ago (R)	0.89	3.55
Fish supply is increasing now because some traders have stopped buying fish (R)	Dropped	

[c] New Scale

Appendix 5.2. Test Results for Discriminant Validity of Constructs

Chi-Square (and CFI) differences between constrained and unconstrained models [a]

	1	2	3
The purchase side			
1 Trust on fishermen	-		
2 Conflict with fishermen	57.22 (-0.17)[b]	-	
3 Dependence on fishermen	51.83 (-0.28)	47.01 (-0.49)	-
4 Supply uncertainty	27.61 (-0.24)	21.91 (-0.31)	25.75 (-0.47)
The supply side			
1 Trust on processors	-		
2 Conflict with processors	104.68 (-0.14)	-	
3 Dependence on processors	19.28 (-0.15)	20.17 (-0.14)	-
4 Supply uncertainty	22.97 (-0.21)	23.56 (-0.15)	24.93 (-0.67)

[a] The critical value ($\Delta X^2 > 3.84$) was exceeded in all tests

[b] Should read as X^2 of the constrained model including trust and conflict is 57.22 higher than the X^2 of the same model when unconstrained. CFI of the constrained model is 0.17 lower than the CFI of the free model.

Appendix 5.3

The discriminant validity tests following Fornell and Larcker, (1981)

Construct	1	2	3	4
1 Trust on fishermen	**.644**[a]			
2 Conflict with fishermen	.314	**.601**		
3 Dependence on fishermen	.134	.026	**.685**	
4 Supply uncertainty	.073	.020	.023	**.774**
1 Trust on processors	**.718**			
2 Conflict with processors	.397	**.659**		
3 Dependence on processors	.325	.084	**.578**	
4 Supply uncertainty	.058	.044	.137	**.746**

[a] The diagonal gives the average variance extracted by the construct and the off diagonal gives the shared variance between constructs. In all the constructs the variance extracted by the constructs is greater than corresponding shared variance.

Appendix 5.4
This appendix gives the distribution of utilities to show how middlemen prefer the different attribute levels.

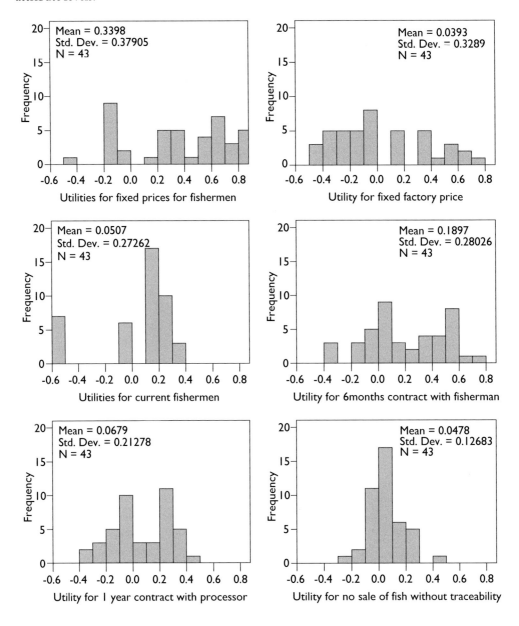

Chapter 6

Implementing sustainable and quality-enhancing practices in the Nile perch channel: the role of the downstream channel members and other stakeholders

6.1 Introduction

The preceding Chapters (3 and 4) identified the market failures that fishermen face to improve welfare and implement sustainable and quality-enhancing practices. The chapters also determined how fishermen address or prefer to address the problems through contracts. Chapter (5) identified double loyalty of middlemen to suppliers (i.e., fishermen) and downstream members (processing factories) as an important factor to implement sustainable and quality-enhancing practices. Although middlemen see the socio-economic problems and are concerned about the long-term availability of Nile perch, they are also constrained by profit considerations. This chapter analyses the downstream partner's involvement in establishing a better balance of the people, profit and planet in the channel (Brundtland, 1987). Specifically, the chapter considers downstream channel members: the local processing factories, importers and retailers. We also consider other stakeholders that are not directly involved in the channel transactions, but are nonetheless important partners to the channel especially in promoting the planet and people dimensions. We refer to this group as the "special interest group" which includes nature conservation and environmental certification organizations, the donor community and non-governmental organizations.

In order to analyse the downstream involvement in establishing a better balance of the PPP issues, we build on corporate social responsibility (CSR) literature (Maignan, Ferrell and Hult, 1999; Prahalad and Hart, 2002). CSR is about the firms' responsibility and action to establish a better balance of the PPP aspects in their activities (Prahalad and Hammond, 2002; Maignan and Ferrell, 2004; Smith, 2003). To-date, CSR literature has predominantly studied the business-society relationship in the context of individual companies (Bhattacharya, Smith and Vogel, 2004; Mackenzie and Podsakoff, 1999; Maignan. and Ferrell, 2003). This chapter extends the analysis to channel–society context where the success of one channel partner's responsibility and actions in balancing the PPP issues may influence or be influenced by other members in other stages of the channel. The chapter therefore undertakes a case study to generate an understanding about how and why different downstream channel members may take responsibility and action to establish a better balance of the PPP issues in the Nile perch channel (Yin, 2003; Eisenhardt, 1989a). Comparable to the structure-conduct performance perspective on business strategy (Porter, 1980), we argue that the position of a firm in the channel determines its conduct regarding its responsibility and actions to establish a better balance of the PPP aspects in the channel. Specifically, the chapter generates propositions about how and why downstream channel members may take responsibility and action to establish a better balance of the PPP issues in the channel.

The rest of the chapter is organised as follows: Section 6.2 outlines the conceptual framework about downstream companies' interests in being involved in solving the welfare, sustainability and quality problems in the upstream. Section 6.3 outlines the study methodology. The methodology outlines the respondents and the case study protocol that was followed to ensure reliability of the results. Section 6.4 presents the results and discussion while section 6.5 gives study implications and prepositions.

6.2 Conceptual framework

CSR literature recognizes two fundamental motivations for firms to take responsibility and actions to establish a better balance of the PPP issues in the total chain, i.e., resource-dependence (Pfeffer and Salancik, 1978) and legitimacy (Suchman, 1995).

The *resource dependency* theory states that "an organization must attend to the demands of those in its environment that provide resources necessary and important for its continued survival" (Pfeffer, 1982, p 193). Stakeholders in each organizations' environment provide different resources which according to Maignan and Ferrell (2004) may vary in importance to the firm. For example, employees provide labour and expertise; shareholders provide capital; customers provide loyalty and regulators can give legal permission to expand facilities. As such, the stakeholder community may have the power to withdraw their resources from the organization if their demands are not attended to (Maignan and Ferrell, 2004). The potential consequences in the event that stakeholders withdraw their support from an organization are what motivate the organizations to attend to the needs of their stakeholders. In that way, organizations protect their access to the resources that the stakeholders control and/or provide to the organization (Hill and Jones, 1992). Following this understanding, fishermen and local upstream communities, as primary custodians of the Nile perch fishery, fall within the environment of those providing resources (i.e., fish supply) to downstream channel members.

Legitimacy is defined as "a generalized perception or assumption that the actions of an entity are desirable, proper or appropriate within some socially constructed systems of norms, values, beliefs and definitions," (Suchman, 1995, p574). In order to be legitimate, a firm should thus align its activities to the predefined norms in the institutional environment in which it operates (Grewal and Dharwadkar, 2002; Suchman, 1995). Deterioration of a channel's legitimacy may send a warning to the channel that it no longer adheres to the norms of expected behaviour and conduct. The legitimacy of the Nile perch channel, for example, has recently been questioned in terms of poverty and food insecurity among local communities around Lake Victoria amidst decades of increased Nile perch exports to lucrative EU markets (see the *Darwin's*

Nightmare documentary, Sauper, 2004). The poverty and increasing food insecurity among local communities signals that the local governments do not protect the vulnerable segments of the society against market failures. It also sends a warning that perhaps the downstream channel members do not equitably reward the primary producers for their contribution to the channel or that they simply do not care about the upstream members. Following these considerations, the downstream channel members are legitimate stakeholders to protect - or help to solve the problems - that upstream channel members face if the channel as a whole is to remain legitimate in the society.

Whereas resource dependence and legitimacy may not be mutually exclusive, their relative importance may differ depending on the position of the channel members in the channel. Currently, there is no coherent theory on CSR in the marketing channels. This chapter generates propositions about why and how downstream members may take responsibility and action to account for the PPP issues in the Nile perch channel by analysing; (1) channel structure, and (2) awareness about -, (3) responsibility over - and (4) willingness to take action to solve the upstream PPP problems. These will be the guiding case study concepts in this study.

6.2.1 Case study concepts

Channel structure: The configuration of the channel and the position of the firm in the channel may have implications for channel relations such as channel power (Stern, El-Ansary and Coughlan, 1996; Coughlan *et al.*, 2001). For example, the concentration of actors in the channel or possession of unique resources may give a channel member power to influence the activities of the other channel members (Heide, 1994). Market power is increasingly becoming evident in the Western developed economies, for example, where supermarket chains are increasingly occupying a powerful position in the food marketing channels. We will briefly examine the structure of the downstream part of the Nile perch channel especially in relation to how it may influence the channel members responsibility and actions to solve upstream PPP problems.

Awareness about upstream problems: Intuitively, one cannot help to solve a problem that one is not aware of. Considering the length of the Nile perch channel some players may be physically disconnected. Consequently, there is no guarantee that all downstream members that may be willing to help to solve upstream PPP problems are well informed about the problems. Intermediaries may benefit from information asymmetry between suppliers and customers (consider price information). Therefore, concealing problems may be in the (short-term) interest of intermediaries or may be a routine behaviour whereby they may deliberately lie or misrepresent the facts to their downstream members (Wathne and Heide, 2000). Further due to information asymmetries, the downstream firms may fail to detect opportunism in their upstream suppliers (Kirmani and Rao, 2000; Wathne and Heide, 2000). This, in turn, may give the intermediaries the opportunity to pursue opportunistic actions without being detected. As upstream problems could be numerous, this study focuses on the PPP problems (i.e., welfare, quality and sustainability) in the Nile perch channel. This study therefore investigates the extent to which the downstream channel members are aware of upstream PPP problems.

Responsibility over upstream problems: Being aware of the upstream problems is one thing, taking or accepting responsibility over the problems is another thing. This study analyses whether or not the downstream firms take responsibility over upstream problems. This is important considering that the downstream and upstream parts of the channels are not mutually exclusive, what happens in the downstream may drive what happens in the upstream and also vice versa.

Willingness to help to solve upstream problems: Accepting (or denying) responsibility over upstream problems may not necessarily translate into willingness to help to solve the problems. It may not always be the case that one only helps to solve problems that one feels responsible for because different people may have different motivations for helping other people. For example, while other people may feel morally responsible to help some people, others may only help if there are mutual benefits attached to their help. This study investigates whether or not downstream channel members would be willing to help to solve upstream problems irrespective of whether they are responsible for the problems or not.

The downstream firms' involvement in solving upstream PPP problems might be beneficial both for the upstream parties and the firms themselves. CSR literature shows that engagement in CSR activities may improve economic performance for the companies (Bhattacharya and Sen, 2004; Maignan, Ferrell and Hult, 1999). In case of enhanced sustainability and quality, local communities will have access to fish for food while the downstream will have long-term fish supply for their businesses. However, such benefits may not always be obvious. For example, Sen, Bhattacharya and Korschun (2006) argue that the impact of CSR in the real world is less pervasive than so far acknowledged because its impact on company's performance may depend on other factors such as quality of the company's products, the type of CSR activities and stakeholder attributions about the genuineness of the company's motives (Sen and Bhattacharya, 2001). Hence, companies may be involved in activities to secure solutions to the PPP problems only to the extent that their involvement fits into their strategies or at least present an opportunity to create shared value for both the society and the company (Porter and Kramer, 2006). So, companies may set conditions for their participation in the activities that promote a better balance of the PPP issues to secure their economic gains. This study investigates the conditions that may motivate the downstream channel members to actually help to solve upstream PPP problems.

6.3 Study methodology

6.3.1 Sources of data

Before data collection commenced, results from the upstream, i.e., from fishermen and middlemen, were analysed to identify major problems and opportunities to improve sustainability and quality in the upstream. For data collection, triangulation was applied, i.e., use of different sources of information (Yin, 2003; Eisenhardt, 1989a). We analysed websites and information fliers to familiarise ourselves with CSR activities in the fisheries sector. Also websites of a number of environmental certification organizations were reviewed. In addition to these sources of data, interviews were conducted with local processing factories and exporters, importers, retailers and special interest groups. During the interviews, further relevant reports and information fliers were also collected from the respondents.

6.3.2 Respondents

The study involved consecutive stages of the channel from local processing factories in Kenya to retailers in the Netherlands. The Netherlands was chosen because of ease of access, budgetary and time constraints to reach out to importers in other parts of the EU. More importantly, the Netherlands is one of the major importing countries for Nile perch in the EU (see www. globefish.org).

Respondents were selected on prior knowledge that they are involved in the Nile perch channel. Interviewing respondents that were already involved in the Nile perch channel was necessary to enable us to address issues specific to the Nile perch channel which other actors may not have been aware of. In the end, we interviewed local processing factories and exporters in Kenya, importers and retailers in the Netherlands and special interest groups[8].

Local processors and exporters: Four local processing factories were interviewed through factory managers and quality controllers. One of the factories had been processing Nile perch for 5 years. Another one has been processing Nile perch for 13 years while two factories have been processing Nile perch for 14 years. The four factories had an installed processing capacity of 15, 25, 30 and 40 tons/day. In the preceding year to the study, these factories utilized between 30 to 60% of their installed processing capacity. These processing factories are members of the fish exporters association of which the executive officer was interviewed.

Importers: Two importers in the Netherlands were interviewed through a plant manager and a managing director. The two importers are among the major Nile perch importers in the Netherlands. Both importers have been importing fish from Lake Victoria since 1990s when Nile perch exports started. Hence, they have good knowledge of the Nile perch export channel. One importer imports Nile perch from the three bordering countries while the other one imports mainly from Tanzania.

Retailers: We interviewed one major retailer and one relatively small, fish specialty shop both also in the Netherlands. Both have been trading in Nile perch products since 1990s. At the

---

[8] We refer to companies interviewed by their generic names as local processors, importers or retailers and special interest groups for those not directly involved in the channel. However, where necessary the names of the organizations will be given. This is done to ensure uniformity of some of the respondents who requested anonymity

major retailer, we interviewed a quality manager while at the fish specialty shop, we interviewed the owner. Although consumers constitute an important part of the channel, they were not involved in the study because we were faced with time and budgetary constraints.

Special interest groups: We interviewed the World Wildlife Fund (WWF) in the Netherlands through the Head of Oceans and Coasts program. The WWF is one of the international nature conservation organizations actively involved in promoting sustainable fisheries through direct support to local communities to engage in environmental protection activities, and through lobbying for policy changes or through public awareness campaigns (www.wwf.org). We also interviewed the Dutch Fish Bureau (www.dutchfish.nl) through the Director. The Fish Bureau provides statistics and databases of the Dutch fish industry, fish exporters and importers among other information. It also gets involved in promoting sustainable fisheries through lobbying for policy change as well as running awareness campaigns among various stakeholders and promoting environmental certification (www.dutchfish.nl).

Other special interest groups were identified through an international workshop on the feasibility of eco-labelling of Lake Victoria fisheries in Nairobi (4-6th October, 2006). The workshop attracted over 40 participants from various public institutions, non-governmental organizations, business sector, research institutions, donor community e.g., the GTZ (Deutsche Gesellschaft für Technische Zusammenarbeit), nature conservation organizations such as WWF (Tanzania) and the World Conservation Union (IUCN) (www.iucn.org), the environmental certification organizations such as the Marine Stewardship Council (MSC) (www.msc.org) and Naturland (www.naturland.com), fishing processing and exporters, the umbrella body for the Lake Victoria fisheries - the Lake Victoria Fisheries Organizations (LVFO). Besides the general workshop discussions that centred on how to promote sustainable practices in Lake Victoria, we had private discussions with a number of participants such as from WWF (Tanzania), Naturland, non-governmental organizations and LVFO.

6.3.3 Case study protocol

Advance preparation, i.e., having a summary of the preliminary results from fishermen and middlemen, and some background information about the CSR activities in the fisheries sector enabled us to develop guiding questions that were appropriate for the respondents in their respective positions in the channel as well as for the special interest groups. In general, however three underlying questions were used to guide the interviews in line with our case study concepts, i.e.:

- The extent to which respondents were aware of the upstream problems.
- The extent to which respondents feel responsible for the problems.
- The extent to which respondents were willing to help to solve the problems.

The interviews were semi-structured to allow flexibility towards specific respondents and two-way discussion with respondents while maintaining focus on the major issues to be discussed (Stewart and Shamdasani, 1990). The interviews started with a basic introduction of the research objectives before specific issues about welfare, quality and sustainability were discussed (Yin, 2003). The opening questions varied across respondents reflecting their positions in the channel. For example, local processors were asked about their perception about the sustainability of the Nile perch and the quality of fish they buy from the middlemen. The importers and retailers were asked about their experiences with the Nile perch products especially in terms of supply quality and quantity. The special interest groups were asked if they had been involved with Nile perch fisheries and if they were aware of any problems associated with Nile perch fishery and channel.

After an introductory discussion, we were able to know the extent to which respondents were aware of specific problems regarding welfare, quality and sustainability and other marketing related problems that fishermen face in their transactions. That helped us to explain the situation in the upstream depending on how much the respondent knew already. Following the description of the upstream situation, a discussion followed on who and why the respondents thought to be responsible for the problems. This discussion enabled us to understand whether the firms take or shift responsibility over the problems. This discussion culminated into the last phase of the discussion about whether or not respondents were willing to help to solve the problems irrespective of who was responsible for the problems. This discussion enabled us to identify the boundary conditions which respondents felt were crucial for the successful solution to the problems.

The interviews in the Netherlands were conducted by two researchers, both taking notes and asking the questions. Interviews with processing factories were conducted by a researcher and a research assistant. Interviews were transcribed immediately after the interviews for further processing and issues requiring clarification were verified with respondents through repeat visits especially with the local processors or email or telephone communication as well as further browsing on websites where necessary for the importers, retailers and special interest groups. Two local processing factories were revisited and a couple of e-mails were exchanged with others. Each personal interview, except those that were conducted at a workshop, lasted for about an hour which is in line with Yin (2003) who suggests that interviews may take about 1 hour. Interviews at the workshop were conducted before or after workshop sessions.

6.3.4 Data analysis

On the basis of the interviews and information retrieved from desk research, a summary report was compiled highlighting the main issues covered. The summary report was used in the analysis about the level of awareness about the upstream problems, responsibility and willingness to help to solve the problems. We compared the results across the stakeholder groups, i.e., the channel members and the special interest groups and along the three main case study concepts. By comparing the results across respondents particularly those actually involved in the channel, we managed to isolate factors that may explain why and how downstream channel members take responsibility and action to solve the upstream PPP problems (Eisenhardt, 1989a).

6.4 Results and discussions

6.4.1 Channel structure
In all, there are 35 processing factories in Lake Victoria region, i.e., 16 in Uganda, 13 in Tanzania and 6 in Kenya. These factories export fish to different parts of the EU. The EU takes up about 80% of Nile perch exports by volume (www.globefish.org). There are about 23 fish importers and exporters in the Netherlands that are involved with Nile perch products (www.dutchfish.nl).

In Europe, most of the food products are traded through supermarkets. In the Netherlands, for example, about 75% of the total food sales is accounted for by supermarkets and of these, over 80% is accounted for by four major supermarkets (Roex and Miele, ed., 2005). In 2005, about 54% of fish products were traded through supermarkets, 25% through the fish specialty markets and 21% through other marketing channels such as the open food markets (see www.hbd.nl). Consequently, the large market share for the supermarkets gives them market power in the food marketing sector to set and enforce standards on their suppliers. The retailers' structural power is further enhanced by the fact that the majority of the products pass through a small number of buying desks. For example, there are 7 buying desks for all the major supermarkets in the Netherlands (Roex and Miele, ed. 2005). This means that suppliers and manufactures may not easily avoid enforcing standards by changing retailers because they would meet the same buying persons for the different supermarkets. With such structural power in the food marketing sector, retailers manage to impose their standards such as the EurepGap on suppliers.

6.4.2 Awareness about upstream problems
The results show that the level of awareness diminishes as we move from local processing factories down to retailers. It should be noted that our study coincided with media attention on the Nile perch channel due to the Academy Award - (Oscar) nominated documentary *Darwin's Nightmare*. The media attention contributed to the level of awareness about the upstream socio-economic problems downstream (see Table 6.1).

Local processing factories and exporters: Processing factories were fully aware of the PPP problems and the trend of events leading to the current state of affairs. This was largely expected given their proximity to the local communities and that they have been operating in the region for some years and experience serious supply problems. They have also suffered from export bans, for example, imposed by the EU in mid 1990's due to quality and safety concerns. Furthermore, they currently fail to process adequate amount of fish due to inadequate supply.

Importers: Both importers were aware of the degradation of the fisheries. It is understandable because they have been involved in the Nile perch chain for some years. One importer was also aware of the poverty, food insecurity and how local traders rely on the factory by-products as well as lack of quality-enhancing facilities in the landing sites. This importer, who once

Table 6.1. Results on the downstream firm's awareness about upstream problems.

Issues	People (socio-economic problems)	Profit (quality improvement)	Planet (degradation of Nile perch)	Remarks
Awareness- to what extent are respondents aware of upstream problems				
Local processors and exporters	• Fully aware • They live in the area. They sell factory by-products to local traders	• Fully aware • They buy fish from middlemen who buy from landing sites	• Fully aware • They fail to operate at full capacity due to inadequate supply	• Due to their proximity to the local communities and also having worked in the channel for long, all processors were fully aware of the problems.
Importers	• One importer was fully aware of socio-economic problems. • Has been to the region several times and seen local traders buying and selling factory by-products • Another importer was not aware of socio-economic problems	• One importer was aware of lack of cooling facilities in landing sites • He has been to some sites personally. • Another importer was not aware of quality problems.	• Both importers were aware of declining fish production. • One importer was optimistic that production can recover within 3-5 years. He said production increased in late 1990's when EU banned Nile perch imports.	• Another importer was only aware of the socio-economic problems after the Darwin's Nightmare. Also learnt about lack of quality facilities in landing sites after our brief description of the upstream situation.
Retailers	• Not aware	• Not aware	• Not aware except for the general degradation of the fisheries resources	• Both retailers were not aware of upstream problems. • Major retailer only became aware of socio-economic problems after seeing *Darwin's Nightmare* documentary.
Special interest groups	• Those not working in the region were not aware of the problems. • Special interest groups working in the region were fully aware of the problems	• Considered to be out of their interest, hence not much discussion over quality issues	• Those not working in the region were not aware of the extent of degradation • Special interest groups working in the region fully aware of degradation	• WWF aware of the declining production. But not the extent of the degradation. • Listed Nile perch on the potentially unsustainable fisheries in their campaign to sensitize European consumers and retailers to demand sustainable fish (see Viswijzer on www.wwf.org)

operated from the region, was among the first to export Nile perch in 1980s. The other importer however was only aware of the poverty, food insecurity and reliance on factory by-products by the local communities after he watched the *Darwin's Nightmare*. It is "*disgusting and not acceptable for the channel and the respective government to let the people starve and yet export everything. Where does all the export revenue go if not to improve the conditions of the local communities,*" he wondered.

Retailers: The fish specialty shop was not aware of any problem with Nile perch channel except for the EU import bans that were imposed in late 1990's due to safety problems. As a result of this ban, the shop sells Nile perch as "Sea perch" for fear that some customers may not buy if they realize that it was the same fish that was once banned. The major retailer was not aware of the socio-economic problems. Our interviews with the quality manager revealed that lack of awareness of upstream problems could be due to misinformation in the channel.

For example, in 2005, the retailer asked a foreign supplier (i.e., an importer from another EU country) for information pertaining to the source of fish, including compliance with food safety standards like HACCP and socio-economic issues surrounding the upstream part of the channel. The importer went to Lake Victoria to collect information from the exporters as requested. In return, the importer reported that, "everything was under control"; this implied that factories were complying with HACCP, and that many people were employed in the factories and fishing sectors, and that their lives were continuing to improve. Shortly after receiving such feedback, the retailer watched the *Darwin's Nightmare* documentary, which showed extreme poverty in precisely the same area where the importer had earlier reported differently. The example suggests that either the importer was protecting his image and business relationship with the retailer by concealing the problems in the event that the retailer would react negatively, and/or the importer was misinformed by the exporters who themselves want to protect their business, or choose to ignore what happens to the local communities even when they see it. This raises questions about information sharing in the channel.

In short, if it were not for media attention of the channel, the major retailer relying on information from it suppliers may not have been aware of the socio-economic problems. This raises the question about the integration of the channel.

Special interest groups: The special interest groups interviewed in the Netherlands as well as some that were interviewed at the workshop in Kenya were not fully aware of the problems surrounding the Nile perch channels. Such respondents heard about the problems in the Nile perch from the workshop presentations. This is because they were not directly involved with Nile perch and they had also come from Europe. However, some special interest groups that are already working in the regions such as WWF (Tanzania), IUCN and the GTZ were already aware of the problems.

In short, the diminishing level of awareness as we move farther downstream suggests that where many independent intermediaries are involved, it may be difficult to make all channel members aware of the upstream problems. Furthermore, dealing with members might also be difficult, particularly if they are thousands of miles apart. Hence firms that are farther downstream may find it worthwhile to take a proactive approach to seek more information about what is going on in the upstream to minimise being taken unaware when problems surface (e.g., through the media).

6.4.3 Responsibility over upstream problems

Local processors and exporters: Local processors and exporters show mixed signals as to their responsibility over the problems (see Table 6.2).

Although the government is responsible for protecting public interests, one manager believes the processing factories are part of the problem and also part of the solution. "*I do not entirely blame the government, to be honest, we cannot be deliberately buying undersize fish and blame the government for not stopping fishermen from using bad gears*", noted the manager when commenting on the increasing use of bad fishing gears and declining Nile perch. All processing factories noted that the government is responsible for providing necessary facilities at the landing sites for good fish handling practices. For their part, they noted that they provide refrigerated trucks to collect fish.

Importers: Although the importers are in the Netherlands, they too blame the local governments for not protecting the poor and enforcing regulations for the sustainable fishing. One importer, for example, noted that the government should protect the poor communities by enforcing policy interventions to ensure food security. The importer also noted that the EU, for example, focuses only food safety issues but not sustainability. The other importer, who has been to the region many times and believes that he knows the behaviour of the fishermen well, blames them for the problems in the communities. He noted that "Fishermen, do not manage the proceeds of the catch very well. They tend to spend too much money on alcoholic beverages and prostitution". The importer also reported that "middlemen force processors to buy also undersize fish together with fish of proper size", which he claims forces the processing factories to process and export undersize fish. As an importer, he noted that he is responsible for the degradation because he accepts juvenile fish. He further blamed lack of cooperation between the bordering governments and corrupt practices in the public institutions that affect enforcement activities in the Lake.

Retailers: The major retailer noted that although as a company they only deal with very limited amount of Nile perch, they feel responsible because as a corporate organization, they would not want to see people who provide them with resources suffer. However, the quality manager noted that he had not been aware of the grave problems in the upstream.

Table 6.2. Results on the downstream members' responsibility over upstream problems.

Issues	People (socio-economic problems)	Profit (quality improvement)	Planet (degradation of Nile perch)	Remarks
Responsibility:- to what extent respondents feel responsible over upstream problems				
Local processors and exporters	• Others feel governments more responsible • Others blame fishermen for not saving and investing the money from fish sales for improving families	• Others feel Government should provide facilities at landing sites • Others feel they already provide refrigerated trucks to collect fish	• All blame government for failing to enforce sustainable fishing • Others blame the processing industry that contributes to unsustainable fishing by buying and processing undersized fish • Also blame fishermen for forcing middlemen to buy undersized fish	Shared responsibility with others like government and also fishermen
Importers	• Feel responsible • Also blame local governments for failing to protect poor people • Also blame fishermen for using money for drinking beers, on prostitution and not helping families	• Not so responsible • Instead they feel local processors and governments should provide necessary facilities in landing sites	• Accept responsibility because they accept juvenile fish in their fish imports from the region. • But also blame EU regulations that enforce quality and safety standards but not sustainability. • Also blame local governments for not enforcing responsible fishing • Also blame lack of cooperation among bordering governments • Also blame fishermen for forcing processors to buy juvenile fish	Shared responsibility with others like government and also fishermen
Retailers	Responsibility as a corporate organization	• Feels responsible as corporate organization	Feels responsible as part of the channel	Feels responsible as a corporate organisation to help to solve problems affecting their upstream suppliers
Special interest groups	Not responsible	Not responsible	Not responsible for the problems	Feels responsible to help to solve the problems

The special interest groups: Although this group is not directly responsible for problems, they noted that as nature conservation organizations they feel responsible to help to solve the problems. Understandably, they are not responsible for contributing to the problems because they are not involved in the daily transactions of the channel.

In summary, it is evident from the results that the responsibility over the problems in the upstream is for all channel members, i.e., local processing factories, importers and retailers. It is also the responsibility of the special interest groups and public institutions. The challenge, however, is who and how to take the initiative to change the situation.

6.4.4 Willingness to help to solve upstream problems

Table 6.3 shows that the respondents' willingness to help to solve upstream problems and some of the prerequisite conditions that respondents feel are necessary for a successful solution to the upstream problems.

Local processors and exporters: Although the local processing factories were fully aware of the PPP problems in the upstream and accept responsibility, they had mixed reactions about whether they should directly be involved in solving problems of fish degradation. Two factories, for example, noted that they were willing to support programs that may solve the socio-economic and ecological problems in the channel. One manager however, was categorical that *"it is the responsibility of the government to stop bad fishing in the lake; we cannot monitor fishermen"*. A quality controller of another factory however had a different view noting that it is very important for the processing factories to be involved in promoting sustainable and quality-enhancing practices in the landing sites. *"We always buy Nile perch of recommended sizes which is the best way for us to contribute to sustainable fishing."*

Although not all processing factories may immediately be willing to be directly involved in promoting sustainable practices, one managing director noted that *"as long as more factories do the right thing, the few that may not be willing, will be compelled to do what the rest are doing."* This illustrates that while some local processors may take a proactive approach, others may be responsive to stakeholder pressure to promote sustainable and quality-enhancing practices.

In short, the results suggest that processing factories look for a level playing field where they should all be involved in addressing the upstream problems. Nonetheless, the local processing factories, on the one hand, fully depend on Nile perch because at their position, they may not easily source raw materials from other suppliers from outside the Lake Victoria region. In addition, Nile perch is currently the only fish that is processed for export markets which means that processing factories risk closing their factories if Nile perch production continues to decline (as some have already closed down – see Chapter 3). On the other hand, in the wake of the negative publicity of the channel, for example, from the *Darwin's nightmare* documentary, the processing factories may also be responding to stakeholder pressure to do something about the problems within their vicinity to improve the image of the channel.

Table 6.3. Results on the downstream firm's willingness and boundary conditions to help solve the upstream problems.

Respondents	Willingness to help solve upstream problems	Important conditions	Remarks
Willingness to help to solve the problems: to what extent are respondents willing to help to solve the problems			
Local processors and exporters	• Some processors were willing to help to solve the problems • Some buy recommended fish sizes to promote sustainability • Can transact directly with fishermen to offer them better incentives like price	• Government commitment to enforce responsible behaviour among fishermen • Level playing field i.e., all processors should implement sustainable and quality enhancing practices • Downstream markets should pay for sustainable products	• Some processors are willing to transact with fishermen directly in which they may get better prices. But fishermen may need to be organized into groups in order to minimise transaction costs
Importers	• Were willing to help to solve upstream problems • May help but put responsibility at local processors and government	• Solving sustainability problems should be market driven • Retailers and consumers should generate demand for sustainable products • Consumers should not only be willing but actually pay for sustainable products	• Believe that Special interest groups like WWF or Green Peace should put more pressure on retailers and consumers to demand sustainable products • May respond to pressure to correct image from negative publicity by Darwin's Nightmare
Retailers	• Willing to help to solve upstream problems but do not know how exactly they could do that	• No conditions attached but need more information about the problems	• Prefer to work with other organizations who may be directly involved on the ground because they may not be physically involved on the ground
Special interest groups	• Willing to help because it is their mandate and policy to protect the environment	• No conditions attached	• But they may work better if there are local institutions in place to enable accountability and enforcement for the sustainable practices

Importers: Both importers were willing to help to solve the problems of Nile perch degradation not only for the benefit of the local communities and fishermen, but also for themselves. As one importer noted, with increasing media attention given to the channel *"we have had to answer many calls from our customers from all over Europe questioning what is happening in the channel."* Both importers however noted that eco-labeling could be a long- term solution for sustainability problems in the Nile perch. However, they both emphasized that the whole idea of promoting sustainable practices should be "market-driven", implying that there should be sustainable market demand for certified fish products. One importer noted that *"there is need to control the retailers and consumers to create demand for sustainable fish products".* The importer further acknowledged that special interest groups like Green Peace and WWF "help" to improve the awareness among retailers and noted that such groups should in fact "continue to put pressure on retailers" to demand sustainable products.

Although both importers noted that demand for certified products has steadily been growing over the past few years, one importer noted that consumers do not always pay for certified products especially if alternative fish is readily available and cheaper. He then noted that importers may easily be compelled to demand sustainable products from their suppliers if *"retailers or consumers are willing to pay for the same."* The importer noted that at the moment the benefits from certification do not always justify the costs because he argued that *"price margins between certified and uncertified fish is not large enough".* In this case, he suggested that in order to motivate more consumers to buy sustainable products, there is need to demonstrate to them for example "if premium prices for sustainability really translates into sustainability on the ground". This, he emphasized, is lacking in the current market information about certification and which may motivate more consumers to buy certified products. For example, it is indicated on the information fliers of one of the importers that the Nile perch channel is "controlled from the lake". While this could assure customers that everything in the channel is under control, the importer however acknowledged during the interviews that they only offer logistical support from the factories.

The importer's argument that promoting sustainable practices should be market-driven could be seen from two perspectives. First, importers may not necessarily depend on Nile perch because they may easily source fish from other sources around the globe. Consequently, they may not be committed to promote sustainable products if there is insufficient demand for it. As such, importers' willingness to help to solve the upstream problems might be due to pressure from the negative publicity of the channel. The worst case scenario would be that importers may abandon the Nile perch channel and start importing fish from other sources for which there might be less stakeholder pressure. As Maignan, Ferrell and Hult (1999, p456) note, *"a responsive organization may choose to address social pressures by moving to a less demanding environment or by altering social expectations through activities such as lobbying."* One of the importers, for example, already noted that the company is becoming less dependent on wild fish because supply from aquaculture production facilities, such as the Pangasius from the Mekong delta region is growing very fast and it is cheaper. Secondly, it may also be a question of channel power. For example, in their position as intermediaries and being less powerful than retailers, the importers may be right that they may easily respond to pressure from retailers than for them to put pressure on the retailers.

However, the importer suggested that the best way for processors, importers and retailers to help to solve problems in the Nile perch channel could be to contribute to charitable organizations such as UNICEF to run CSR programs to alleviate poverty or to nature conservation organizations such as Green Peace or WWF or others which could run micro-projects among fishermen to enable them to access recommended fishing gears for sustainable fishing and tools for improving quality.

Retailers: Despite the limited information that the retailers had about upstream problems, retailers generally were willing to help. The fish specialty shop owner, for example, noted that if the channel can certify with MSC, he would be willing to sell certified fish which he believes would easily sell in his local market.

The major retailer was also willing to help to solve the upstream problems but indicated that it was not certain how to help and that more information was needed. Although the quality manager admitted that Nile perch is not a significant proportion of the fish products, he noted that once he gets adequate information, he would take up the issue of Nile perch with the board of the company and other major stakeholders such as those involved in the EurepGap to find out how they can help in solving the problems. The retailer posted a similar response on its website in response to the *Darwin's Nightmare documentary* indicating that it will not stop buying Nile perch but instead it would work with other partners to improve the situation. He also noted that since as a company they are some stages away from the primary stages, it would be difficult for them to be physically involved in the primary stages but instead, he would be willing to work with special interest groups that may directly work with fishermen and local communities.

The willingness of the retailers to help to solve the problems may not be unexpected. Although the retailers might not necessarily depend on Nile perch fish, they may need to protect their image. In view of their visibility in the food marketing channels, major retailers may easily be targeted by the environmental campaigners and may be extremely vulnerable to negative publicity. Consequently, retail chains try to listen and watch market signals closely and then react efficiently. Alternatively, the retailer may use its power in the channel to influence change on their suppliers.

Special interest groups: The special interest groups were willing to help to solve the upstream problems. For example, the WWF (Netherlands) noted that as a nature conservation organization, they are overly concerned about the degradation of the fisheries around the globe and that they would be willing to help to solve the problems. The respondent noted that like in other places around the world, *"funding for sustainable fisheries programs for poor communities would not be a problem.*" WWF helps local institutions and communities to source funds to promote environmental programs such as sustainable fisheries. Although WWF has not yet been involved with Nile perch, he noted that helping the local communities in Lake Victoria would be in line with their mandate and strategy to promote sustainable fisheries. However, he noted that there is need for local institutions and clear standards for sustainable practices such as defining fishing zones, fish sizes and type of fishing gears. To that, he was pleased to note that local institutions composed of fishermen and local leaders within the villages already exist that implement sustainable practices and that standards for implementing sustainable fisheries already exist in the legal frameworks in the region.

Other special interest groups also expressed strong willingness to support sustainable practices in Lake Victoria. For example, in the wake of the need for eco-labeling of the Lake Victoria fisheries, MSC and Naturland – both certifying organizations noted that they would support the certification process. Naturland, for example, noted that they would be willing to start pilot projects with selected sites and groups of fishermen where they could start to monitor sustainable practices pending possible certification. There is also willingness in the donor community to improve both sustainability and livelihoods of the small-scale fishermen and fishing communities. For example, starting with the workshop on eco-labeling of Lake Victoria fisheries, GTZ together with MSC and Naturland are planning pilot eco-labeling projects in the three countries bordering the lake.

In short, special interest groups such as the WWF and certification organizations might be willing to help solve the problems of fisheries degradation as this is why the organizations were established in the first place. In the case of the donor community, it is part of their mission to help to alleviate poverty in developing economies. Nonetheless, their willingness to help to solve the problems in the Nile perch signals an opportunity to approach the problems from both ends, i.e., putting pressure in the downstream and also supporting upstream primary producers to implement welfare, sustainable and quality-enhancing practices. That way, the intermediaries such as the importers and retailers may be compelled to do something about upstream welfare, sustainability and quality-enhancing practices.

6.5 Study implications and propositions

The objective of this chapter was to generate an understanding about how and why different downstream members in the channel may take responsibility and action to establish a better balance of the PPP issues in the Nile perch channel. This section discusses the implications of the results and generates propositions about the level of awareness, responsibility and willingness to take action to solve the PPP problems for the different members in the different positions in the channel.

The results show that as the channel moves further downstream from local processors to the retailers, dependence on Nile perch declines. This is the case because in the downstream, Nile perch as a commodity may easily be substituted with other fish products such that if there are any problems with it, downstream members may easily switch to other fish products from other sources around the globe. This is evidenced by the fact that one importer acknowledged that he is becoming less reliant on wild fisheries as alternative sources from aquaculture become readily available and even cheaper. It was also evidenced by the major retailer who acknowledged that Nile perch constitutes a very small portion of their fish products. Consequently, while importers and retailers may be concerned about fisheries degradation in the global supply chain, the degradation of Nile perch may not significantly affect their businesses. In the upstream however, the Nile perch may be less substitutable. In the Nile perch channel, for example, local processing factories only process Nile perch and those that fail to source adequate supply from Lake Victoria due to declining production either close down or operate at low capacity with unlikely possibility to source from other sources in the region or other sources around the globe.

As Porter and Kramer (2006) argue, companies may prioritize the social issues in their external environment that significantly affect the company's competitiveness or are significantly affected by the company's activities in the course of business. Where both of these do not apply, the competitiveness of channel members farther downstream may not be significantly affected by declining channel-specific resources such as Nile perch in our case. So we propose that:

P1: The farther downstream a company is in the channel, the less dependent it is on upstream substitutable channel-specific resources

However, low resource-dependence on specific channel resources may not necessarily imply that downstream firms may not be responsible for the upstream problems. Channel members and other stakeholders may be involved in establishing a better balance of the PPP aspects because they want to create benefits for others, or mutual benefits and/or reduce losses for themselves. For example, the special interest groups, who are not necessarily involved in day to day channel transactions, are willing to be involved to solve the upstream PPP problems in the Nile perch channel for the benefit of the local communities and the sustainability of the fisheries. The local processing factories and importers may be involved in order to create mutual benefits for their business as well as for the communities. Perhaps, that is why they seek joint responsibility over solving upstream problems. The retailers may be involved in solving upstream problems to reduce losses that they may incur especially if special interest groups mount negative publicity about the PPP problems in the channels that may damage the retailer's image. The negative publicity may create more losses for the retailers than the actual upstream PPP problems. Further, the retailers are willing to work with special interest groups. In that way, they may also minimise costs of being directly involved in dealing with primary producers and local communities. They may also prevent further negative publicity by the special interest groups that they would work with.

Whatever the motivation, the results show that downstream channel members are more responsive than being proactive in their willingness to establish a better balance of the PPP issues in the channel. This responsive approach can be understood from the fact that although Nile perch may easily be substituted by other fish supply from other sources, downstream channel members may still be held responsible for the problems in the Nile perch channel. Hence, the responsibility and actions of the channel members that are farther downstream to help to solve the upstream PPP problems may not be driven by their dependence on the Nile perch but by legitimacy pressure from stakeholders. Thus, we propose that:

P2: The farther downstream (upstream) a channel partner is in the channel of a substitutable resource, the more its responsibility over specific upstream PPP problems will be motivated by legitimacy (resource-dependence) arguments.

Since firms that are farther downstream may be less dependent on Nile perch, they may not be aggressive or proactive enough to source information about specific upstream PPP issues in the channel. The retailers, for example, were not aware of the PPP problems in the upstream of the channel. Lack of awareness by the retailers shows that the channel is weakly integrated such that importers and exporters operate as independent intermediaries. It may also come down to the fact that the retailers at their position feel detached from the causes and consequences of the specific upstream PPP problems in the Nile perch channel. Hence, in absence of other sources of information such as media or other stakeholders, channel members farther downstream are less likely to be aware of the PPP problems associated with upstream substitutable channel-specific resources. We thus propose that;

P3: The less dependent a downstream company is on channel-specific resources, the lower the awareness about upstream channel specific PPP problems (other things constant).

Whether downstream channel members are aware of and feel responsible for the upstream PPP problems on the basis of resource dependence or legitimacy arguments, may not necessarily mean that they may easily take action to solve the problems in the channel. Resource dependent channel members may take action anyway to solve, for example, the sustainability problems because their businesses may be more at risk if resources get depleted. For the channel members that feel responsible only on the basis of legitimacy arguments may act only if other members act, i.e., level playing field or if they have sufficient power to influence change. For example, the importers noted that they are willing to solve the PPP problems in the Nile perch channel but that there should be market demand. However, as the importers pointed out, the retailers in their position and the market power that they have may easily influence change in the upstream part of the channel. The retailers may also have appropriate institutions such as the EurepGap to enforce change in the channel. In contrast, intermediaries such as importers and exporters may not have the power or appropriate institutions to enforce their demands on the retailers. Hence channel members that feel responsible for the upstream PPP problems because of legitimacy reasons may be willing to solve the upstream PPP problems in the channel of substitutable resources and more importantly so, if they have sufficient market power to impose the PPP improvements on others. So we propose that:

P4: The more the responsibility of downstream channel members for upstream channel specific PPP problems is based on legitimacy arguments, the more their willingness to take action to solve the PPP problems will depend on whether they have sufficient market power

6.6 Conclusion and future research

In conclusion, this study has developed a channel perspective of the underlying factors that may influence the downstream firms awareness, responsibility and willingness to help to solve upstream PPP problems in the Nile perch channel. These results show that as the channel moves farther downstream, firms' dependence on the Nile perch as well as downstream channel partner's awareness of specific PPP problems diminishes. This means that if channel members farther downstream are going to "help" to establish a better balance of the PPP issues in the channel, it will be based much more on the legitimacy arguments raised by stakeholders and media. Whether the channel members will respond to stakeholder pressure, may also depend on whether they have the market power to force other channel members to act with them in the channels.

These results have implications especially on how to get the downstream channel members involved in solving upstream problems. It may require the special interest groups and other stakeholders to raise awareness and put pressure on the downstream channel members to get them involved in solving upstream problems in the Nile perch channel. The results also show that the firms closer to the upstream (e.g., local processors) were willing to help to solve the problems because of both resource-dependence and legitimacy arguments. While they face declining supply without viable alternative sources of input fish supply, they also face increasing downstream pressure to do something about upstream problems. These results further show that the channel is weakly integrated, i.e., there is a disconnection especially in information flow and sharing between the upstream and downstream. These results therefore support the idea of alternative governance mechanisms such as contracts to oblige channel members to account for the PPP issues in the channel. With contracts, retailers must demand sustainable products from importers who should also demand the same from the exporters. In that way, the chain of reaction would go upstream to the primary producers.

While this study gives extensive insights about how and why the different members in the downstream part of the channel may take responsibility and action to establish a better balance of the PPP issues, it also highlights potential areas for future research. There is need to undertake quantitative research to test our propositions. This may involve considering different channels, for example, perishable and non-perishable products and food and non-food products. Since this study was done on common property fisheries resources, future research may consider testing our propositions on other common property resources such as forestry products.

Chapter 7

Conclusion, discussion and policy implications

7.1 Introduction

This thesis focused on the people, profit and planet (PPP) issues (Brundtland, 1987) in international marketing channels. International marketing channels especially those originating from developing economies, face a dilemma in establishing a better balance of the people (welfare of local communities in developing economies), profit (for all channel members) and planet (degradation of natural resources - NR). We considered all theses issues and how they can be balanced in international channels. Whereas different mechanisms may help to balance the PPP issues in the channel, this thesis emphasized contracts as a way to secure sustainability of NR in a manner that improves welfare and profitability at primary production level. The thesis considered governance structures around the chain step by step by considering different stages and channel members along the channel from primary production to retail distribution. The central question was whether contracts might be used to establish a better balance of the PPP dimensions in international marketing channels and what should be the terms of contract and governance structures that may stimulate welfare, profitability and sustainability at the primary production level and how the downstream channel members can help the channel as a whole to better balance the PPP issues. The thesis addressed the following specific questions:

- What are the market failures that constrain the ability of primary producers to establish a better balance of the PPP issues in their activities?
- What terms of contracts and governance mechanisms can stimulate a better balance of welfare, sustainability and quality at primary production level?
- In what way can downstream channel members and other stakeholders help to stimulate welfare, sustainability and quality in the channel and especially at the primary production level and what would motivate them to do so?

This chapter gives a conclusion of the previous chapters, a general discussion and policy implications. The rest of the chapter is organized as follows. Section 7.2 gives a summary of the conclusions from the major findings. This is followed by section 7.3 that gives a general discussion of the results. Section 7.4 gives policy implications followed by section 7.5 that gives theoretical considerations. Section 7.6 concludes the chapter by highlighting potential areas for future research.

7.2 Conclusion

This study was set out on the presumption that by changing the behaviour of fishermen and other channel members, the situation at Lake Victoria may also change and that contracts can be a mechanism to bring about such behavioural change among fishermen and middlemen. However, deciding how contracts may change the behaviour of the fishermen necessitated an inquiry into why fishermen behave the way they do. We can conclude that fishermen fail to implement sustainable and quality-enhancing practices because of the major bottlenecks that they face. The bottlenecks include the degradation of the NR (i.e., fisheries), access to production facilities, information asymmetries and ineffective enforcement (see Table 7.1 for a summary). Unless these market failures are addressed, it may be difficult to establish a better balance of the PPP issues at primary production level and in the channel as a whole.

On the basis of the results (Chapter 4), we conclude that contracts are acceptable mechanisms to fishermen. In order to change the behaviour and stimulate responsible activities, fishermen prefer sustainability and quality-enhancing contracts that provide production facilities, price information, bring them closer to international channels and allow private policy enforcement of sustainable practices. However, fishermen have idiosyncratic preferences for the terms of the contracts such that they should be offered a choice among different contracts. In order to drive

Table 7.1. Summary of the market failures that fishermen face.

Market failures	Contributing factors	Consequences
Degradation of the fisheries	• Over-fishing • Use of bad fishing gears • Limited sources of livelihood • Ineffective enforcement • Ecological factors	• Food insecurity and malnutrition • Increased threats to sustainability
Limited access to production facilities	• Policy failure to provide credit facilities for small-scale fishermen or make the production facilities affordable for poor fishermen	• Lack of tools for quality improvement • Increased use of relatively cheaper but destructive illegal fishing gears • Fishermen obtains production facilities through informal loans from fish buyers • Create interlocked credit/fish markets • Fishermen lose bargaining power
Information asymmetries	• Lack of marketing institution to control or disseminate market information • Lack of transparent marketing systems (e.g., auctions)	• Fishermen lose bargaining power • Price risks • Income risks
Ineffective enforcement	• Inadequate financial and human resources for monitoring and enforcement • Enforcement biased against fishermen and not fish buyers	• Inadequate incentives for fishermen to implement sustainable practices • Middlemen and processing factories feel less responsible to implement sustainable practices than fishermen

the responsible behaviour among fishermen, middlemen should provide market information and enforce sustainable and quality-enhancing practices on fishermen (Chapter 5).

In order to create a situation in which fishermen and middlemen are engaged in sustainability and quality-enhancing contracts, downstream channel members (such as local processing factories) should either engage into direct transactions with fishermen or impose necessary conditions to ensure that middlemen undertake responsible buying and supplying activities. Channel members farther downstream (such as importers and retailers) and special interest groups may help by creating and/or supporting micro-projects that may enable fishermen to solve the market failures that they face.

7.3 General discussion

This section gives a general discussion of the results of the study. We discuss (1) channel members motivation to establish a better balance of the PPP issues in the channel and (2) the necessity of contracts.

7.3.1 Motivation for establishing a better balance of the PPP issues

The analyses in this thesis demonstrate that channel members support the need to establish a better balance of the PPP issues in the channel despite that they may have different motivations for doing so. The primary producers rely on fisheries for survival and business. Hence, they may be motivated by both prospects of better welfare and also enhanced profitability of their business activities to meet other socio-economic needs. The middlemen may be motivated to balance the PPP dimensions in order to secure their short-term economic interests as well as long-term business continuity which are already constrained by their double loyalty to suppliers and buyers. The downstream channel members may be motivated by their dependency on the resources and stakeholder pressure depending on their position in the channel (either closer to the upstream or farther downstream respectively). Responding to stakeholder pressure is also economically motivated because such channel members may seek to protect other resources that they depend on such as their image and identity which may be at stake in the event of negative publicity by stakeholders such as special interest groups.

Although different channel members support the need to establish a better balance of the PPP dimensions for different reasons, not all feel fully responsible for the upstream PPP problems in the channel. Throughout the channel, different members blame each other as well as other stakeholders like the local governments for the problems. This shared responsibility among channel members and other stakeholders may be attributed to the (common) property rights of the NR (fisheries). Literature on common property resources acknowledges that resource users often do not take responsibility over resource management (Brox, 1990; Demsetz, 1967, Hardin 1968). Channel members in all the stages expressed a need to enforce rules and regulations on all channel members to ensure a level playing field either through public or private policy. It is evident in our analysis that how rules and regulations for sustainable practices are enforced is a motivating factor in the decision to implement sustainable practices at primary production level (Chapter 4). It is also evident in the results (Chapters 3 and 6) that members not only blame the local governments for the problems but also put responsibility for solving the problems on the same local governments.

7.3.2 The necessity for contracts

Since channel members do not feel fully responsible for the PPP problems and that they may have different motivations for establishing a better balance of the PPP aspects, there is a danger that without proper coordination of their activities, channel members may be opportunistic. For example, the primary producers may be opportunistic towards welfare and profit, and because they have pressing short-term socio-economic needs, they may easily use any means to meet their short-term needs even if that may imply emptying the lake (i.e., compromising the planet dimension). This is evident in Chapter 3 where, for example, the use of destructive fishing methods is partly a result of the need to secure daily food needs. The middlemen and downstream channel members may be opportunistic towards profit. This is evident, for example, in that middlemen and local processing factories buy and supply undersized fish in order to enhance their short-term economic gains thereby compromising the people and planet dimensions. The special interest groups may be opportunistic towards the planet and people dimensions because they are interested in conservation of NR and the welfare of local communities, thereby compromising the profit dimension of the chain as a whole.

Therefore, in order to better balance the PPP aspects in the channel, there is a need to manage the opportunistic behaviour among chain members and other stakeholders. Such opportunism can be managed by public or private policy. Public policy can manage opportunism by being more effective in enforcing the rules and regulations for sustainable practices to ensure protection of NR and guarantee a level playing field for all. It is however, evident that public policy, especially the local governments, faces limited human and financial resources to monitor and detect opportunistic behaviour in the channel. Private policy enforcement can be achieved by establishing and enforcing rules of engagement in the channel. For example, in the same way downstream channel members enforce quality control through private standards, they may enforce sustainability in the channel. This can be achieved through contracts between channel partners.

Contracts are also necessitated by the fact that the channel is weakly integrated. The presence of complex market failures at primary production level and lack of awareness about the same in the downstream part of the channel is testimony to that. The implication of weak integration of the channel is that channel information does not easily flow from one stage to another and consequently members in the different stages may easily pursue opportunistic behaviour without being detected. In that case, it may be difficult to manage opportunism and establish a better balance of the PPP issues in the channel. Therefore, weak channel integration enhances the necessity of contracts to oblige channel members to undertake activities that may establish a better balance of the PPP aspects in the channel. Such contracts should thus focus on minimizing information asymmetries, opportunistic behaviour and enhanced quality control in the channel.

On the one hand, minimizing information asymmetries at primary production level may enable primary producers to minimize welfare risks that arise from price and income risks, and degradation of NR. On the other hand, minimizing information asymmetries in the downstream part of the channel may enable channel members especially those farther downstream to detect and minimize opportunistic behaviour among intermediaries. More importantly, channel members that have the power to change the situation in the channel (such as the retailers) may easily make the intermediaries to undertake activities that contribute to the better balance of the PPP aspects in the channel.

7.4 Policy implications

From the preceding discussion, it is apparent that contracts are necessary to stimulate a better balance of the PPP issues at primary production level. However the implementation of the contracts raises both public and private policy implications. This section discusses such policy implications starting with private and then public policy implications.

7.4.1 Private policy implications

In order to address the complexity of the market failures at primary production level, sustainable and quality-enhancing contracts should focus on enabling the primary producers to have objective means to secure their livelihoods, (i.e., stable prices and incomes) and most importantly, production facilities to undertake activities that protect the planet (i.e., appropriate fishing gears) and enhance profitability (i.e., tools for quality improvement).

Primary producers and buyers can engage in contracts that focus on provision of market information, production facilities and also enforce sustainable practices. However, there might be a danger of creating interlocked markets (Bardhan, 1980) if primary producers obtain both market information and production facilities from the buyers to whom they sell their commodities (whether middlemen or local processing factories in this case). The overdependence on buyers for commodity- as well as credit markets for production facilities may compromise the primary producers' bargaining power and be prone to manipulation by buyers (Bardhan, 1980; Glover, 1987; Key and Runsten, 1999). As this is already evident at the primary production stages where fishermen who obtain fishing gears from buyers tend to be locked into relationships without chance to decide where to sell their fish or else they loose access to production facilities (Chapter 3), the approach should be changed. Therefore buying and supplying contracts should be separated from those that provide production facilities. Separating the two means that primary producers and buyers should only engage in *market-specification* contracts in which they focus on marketing issues such as price, quality and delivery schedules (Key and Runsten, 1999; FAO, 2001).

The primary producers should then obtain the much needed production facilities from other sources or engage in separate *resource-providing* contracts with other stakeholders. This can be achieved in different ways. Primary producers may obtain credit from micro-finance institutions to buy production facilities. However, as these facilities are not readily available and where they are available, they often do not target poor people (Basu, 1997; Coleman, 2006), other mechanisms should be considered. For example, the special interest groups that already support rural communities to implement environmental programs should set up micro-projects through which primary producers can obtain production facilities. Setting up micro-projects might be more feasible because the special interest groups are willing to support the sustainable activities at primary level. In addition, some downstream channel members prefer to support such groups instead of themselves being involved at the primary production stage.

However, whether primary producers secure separate *market-specifications* or *resource-providing* contracts, there is a need for group action on the side of the fishermen. Group action may enhance their bargaining power in terms of market prices, coordinate quality control and enforce social control to safeguard groups' interests. As group action often fails due to conflict of interest among members, there may be need to offer different contracts to group members according to their preferred terms of contract. Although there are some groups at primary production stage that are already organized around the need to promote sustainable practices, it may not necessarily imply that they can have the same contracts. While some members may want production facilities, others may only want price information. These preferences need to be honoured even if contracts could be coordinated through group action.

Although downstream channel members are willing to help to solve the PPP problems in the channel, they obviously do not need contracts similar to those at the primary production level. However, there is a need for mechanisms to get them actually involved in solving the PPP problems. Such mechanisms could be sustained pressure from the market and stakeholders such as the special interest groups. The special interest groups and the media should continue to expose problems in the upstream part of the channel and mobilise and lobby consumers to demand sustainable products. The retailers may need to generate demand for sustainable products or demand certified products. Pressure on and also by the retailers may generate a chain of reactions all the way to the upstream part of the channel. Our results show that even the intermediaries recognise the impact that pressure from the market as well as retailers may have on their activities.

7.4.2 Public policy implications
Although contracts are essentially private, their success may still need public policy support. The public policy should undertake pro-poor policies to enable poor primary producers to access production facilities. For example, the public policy should set up such micro-finance institutions that deliberately target poor primary producers.

Public policy should also provide an enabling regulatory environment to enforce sustainable practices as well as an enabling physical environment that provides public infrastructure necessary to undertake some crucial activities. The public policy should enforce rules of sustainable practices on all channel members in order to ensure level playing field and minimise opportunistic behaviour in the channel. In international marketing channels, however, there is a limit to which local governments may enforce rules and regulations on channel members from across the borders. Therefore, there should be collaboration with international institutions such as the European Union (EU) to enforce sustainable practices on imported NR commodities. Such collaboration in enforcing sustainable practices across borders is important to ensure a level playing field between channel members that fall in the jurisdiction of the local governments and international institutions.

Public policy should also provide public infrastructure, goods and services such as power (electricity), running water, cold storage facilities which are essential when handling fresh products. At the primary production stage, such infrastructure and facilities may facilitate storage of fresh commodities in the event that there is less demand or producers are not satisfied with market conditions such as prices. Storage facilities may also enable primary producers to keep their commodities until the scheduled delivery time in the case of contract delivery schedules. Although (fish) production may be done on daily basis, the production may not always warrant daily deliveries because of declining production.

7.5 Theoretical considerations

This thesis is among the first to study sustainability issues across the channel as a whole from primary production to retail distribution and importantly so, for an international channel between developing and developed economies. In addition, due to the limitations in the existing theories of channel governance to guarantee sustainability of natural resources, this thesis combined a number of theoretical perspectives to develop and test theoretical arguments for the use of contracts to stimulate sustainable and quality-enhancing practices at the primary production level. Specifically, this thesis developed a number of theoretical arguments based on the transaction cost economics (TCE), social and network theory, as they may apply to common property resources. The thesis also built on corporate social responsibility (CSR) literature to examine how and why downstream channel members may take responsibility and action to solve the PPP problems in the upstream part of the channel. A number of theoretical arguments raised in this study have been confirmed thus raising new insights about how an integrated theoretical approach can be used in the application and use of contracts to promote sustainable and quality-enhancing practices in the marketing channels.

By applying the different theoretical arguments in favour of contracts to common property resources, this thesis brought new perspectives to the use of contracts that has predominantly focussed at risk minimization, protection of specific assets, information asymmetries and opportunism in profit maximizing transactions following Coase (1937) and subsequently,

Williamson (1985). Our analysis shows that while channel members especially at the primary production stages may seek to minimise long-term risks that may arise from the degradation of NR, they are not intending to promote self interests only. By willing to implement sustainable and quality-enhancing practices and particularly so, in common property resources, primary producers are, in fact, intending to promote public interests as well. Primary producers are also willing to take on risks in that they may not guarantee returns from implementing sustainable practices especially if other resource users do not implement the same practices. For example, by leaving fish in the lake to grow for future use, primary producers may have no reason to believe that they will catch the same fish when it grows because other fishermen especially those that do not practice sustainable fishing may likely catch it. In addition, primary producers are willing to take on welfare risks because as fish production declines, they may sometimes not be able to catch any fish for their own welfare at least in the short-run before the production recovers. Therefore, this thesis shows that while risk-minimization is an important motivation for transaction partners to choose governance structures, it may also depend on the type of risks in question. Hence, the TCE assumptions that agents may always choose governance structure that minimize risks and also that they may be opportunistic, i.e., promote self interests (Williamson, 1985) may not hold in all contexts (Geyskens, Stenkamp and Kumar, 2006) or at least not in the context of this study where at the primary production stages, producers are willing to take on short-term risks in order to minimize long-term risks and promote sustainable practices in the interest of public interest.

Following the TCE arguments, previous studies have demonstrated that primary producers appreciate contracts that minimize such market failures as access to input and output markets, information asymmetries, production facilities and quality control (Masakure and Henson, 2005; Porter and Phillips-Howard, 1997; Saenz-Segura, 2006). This thesis has not only confirmed these arguments but also identified the terms of contracts that may stimulate sustainable and quality-enhancing practices at primary production level under particular contextual factors. To the best of our knowledge, the application of contracts to stimulate sustainability of common property NR is not yet common in marketing literature.

Following the social and network theory, previous studies demonstrate that transaction partners would choose both governance mechanisms and transaction partners depending on the social and network factors in their transaction process (Houston. and Johnson, 2000; Wathne, Biong and Heide, 2001; Wuyts and Geykens, 2005). Previous studies have also argued that social relations and contracts are either complementary (Poppo and Zenger, 2002) or substitutes (Larson, 1992; Uzzi 1997) in the way they coordinate economic transactions. This thesis has confirmed that social relations and network ties are important determinants of the terms of contract such as selection of contract partners and whether to sustain or exit relationships. However, lack of significant effect of social relations on the middlemen's preference for the terms of contracts with processing factories (Chapter 5) raises questions about whether or not social relations may influence the choice of contracts in an oligopsony (i.e., where suppliers are in larger number than the buyers) (Scherer and Ross, 1990), and also where suppliers operate on much smaller scale relative to the buyers. This remains an open question for future research.

Building on CSR literature, this thesis has unravelled insights into how and why downstream channel members may take responsibility and action to balance the PPP issues in the upstream and the channel as a whole. This thesis particularly extended the CSR application from business–society relationship within the context of individual companies to a channel-society perspective where channel actors may have to work together to establish a better balance of the PPP dimensions. Our analysis shows that the downstream channel members may support activities to establish a better balance in the PPP dimensions in the upstream either on the basis of resource dependence and/or legitimacy arguments (i.e., stakeholder pressure) depending on their position in the chain (closer to the upstream or farther downstream).

7.6 Future research

In the different chapters of the thesis, a number of limitations and suggestions for future research have already been brought forward. However, there are general limitations that cut across all chapters and hence they need to be highlighted here. This study has largely dealt with channel members' intentions to balance the PPP issues in the channel. On the basis of these intentions, this thesis concludes that contracts may help to establish a better balance of the PPP issues especially at the primary production level. An important task for future research would be to undertake similar research in other NR both food and non-food commodities such as forestry products, or in the same fisheries but from different geographical settings (such as in Asia) to corroborate these results.

Another point for future research is the scope of the research. Data used in this thesis was collected from Kenya – one administrative part of the Lake Victoria Fisheries and in the Netherlands – one importing country in the EU. Although administratively the Lake is divided into three countries, fish, especially Nile perch known to be highly migratory, sees no geographical boundaries. Hence, its fate or fortune for sustainability hinges on what happens in the three countries. Future research should therefore consider including Tanzania and Uganda in order to establish mechanisms that are applicable and acceptable to the entire Lake region. Similarly, fish from Lake Victoria and even so, from Kenya goes to different parts of the EU as well as outside the EU. Since EU remains a major market for the Nile perch, future research should consider including channel members from different regions and countries in the EU. Such a scope may give better insights into how integrated or otherwise the channel is. Such insights might also help channel members willing to support the balance of the PPP issues to know where, how and when to implement change.

Although this study focused on contracts, it does not necessarily imply that contracts are the only mechanisms that may enhance sustainable practices among small-scale producers. Other mechanisms such as certification should also be considered (Ingenbleek and Meulenberg,

2006). Currently, there is limited participation of small-scale primary producers in certification programs. It is not certain how they can be involved. It would be interesting if future research would investigate how contracts can be used in combination with certification.

Although in our policy implications we have noted the importance of group action for the primary producers, we recognise that group action does not always work. In Kenya, some fishermen groups that are integrated within the local institutions at the beach level are, however, managing to enforce sustainable practices through social control (Kambewa, Ingenbleek and van Tilburg, 2006). Investigating how such groups operate and how they can help to trigger a bottom up approach to solving problems that affect primary producers might be an important aspect of future research.

To crown it all, any research result is meaningless if it does not contribute to solving problems that the research was intended for. And solving the problems is not a function of the results per se but the implementation of the results. The value of this research will depend on how much the implementation of these results will contribute to the People, Profit and Planet balance in Lake Victoria and the channel as a whole. This thesis has paved the way for this and future research could build on it.

References

Aakkula, J., Peltola, J., Maijala, R. and Siikmaki, J. (2005): Consumer Attitudes, Underlying Perceptions and Actions Associated with Food Quality and Safety, Journal of Food Products Marketing, Vol. 11(3), 67-87.

Abila, R.O. (2000): The Development of the Lake Victoria Fishery. A Boon or Bane for food security? Socio-economics of the Lake Victoria Fisheries. IUCN, Report No. 8.

Abila, R.O (1998). Four decades of the Nile Perch Fishery in Lake Victoria. Technological Development, Impacts and Policy Options for Sustainable Utilization. In Matindi G.W. and Matindi, S.W (eds): Water Hyacinth, Nile Perch and Pollution: Issues for Ecosystem management in Lake Victoria. Proceedings of a workshop on Prospectus for sustainable management in Lake Victoria. Mwanza, Tanzania 10-12 June 1998.

Abila, R.O. and Jansen, E.G (1997): From Local to Global Markets. The fish exporting and Fishmeal industries of Lake Victoria - Structure, Strategies and Socio-economic impact in Kenya. Socio-economics of the Lake Victoria Fisheries. IUCN, Report No. 2.

Achrol, R.S. (1991): Evolution of the Marketing Organization: New Forms for Dynamic Environment, Journal of marketing, 55, 77-93.

Achrol, R.S. and Gundlach, G. (1999): Legal and Social Safeguards against opportunism in Exchange. Journal of Retailing, Vol. 75 (1) pp 107-124.

Achrol, R.S. and Kotler, P. (1999): Marketing in the Network Economy. Journal of Marketing, Vol. 63, (Special Issue), 146 -163.

Aggarwal, R.M. (2006): Globalization, local ecosystems, and the rural poor World Development, Vol. 34, No. 8, 1405- 1418, 2006 .

Agrawal, A. and Chhatre, A. (2006): Explaining Success on the Commons: Community forest Governance in the Indian Himalaya, World Development Vol. 34, Issue 1, 149-166.

Aiken, L. and West, S. (1991): Multiple Regression: Testing and Interpreting Interactions. Newbury Park, CA: Sage.

Andaleeb S.S. (1995): Dependence relations and the moderating role of trust: Implications for behavioural intentions in marketing channels. International Journal of Research in Marketing 12, 157 -172.

Anderson, J.C. (1987): An Approach for Confirmatory measurement and Structural equation modelling of organizational properties. Management Science, Vol. 33 No 4, 525 -541.

Angulo, A.M., Gil, J.M. and Tamburo, L. (2005): Food Safety and Consumers' Willingness to Pay for Labelled Beef in Spain. Journal of Food Products Marketing, Vol. 11(3), 89-105.

Antia, K.D and Frazier, G.L (2001): The Severity of Contract Enforcement in Interfirm Channel Relationships: Journal of Marketing, Vol. 65, 67-81.

Argyres, N.S. and Liebeskind, J.P. (1999): Contractual commitments, bargaining power, and governance inseparability: Incorporating history into transaction cost theory. Academy of Management Review, 24, 49-63.

Atuahene-Gima, K and Li, H. (2002): When Does Trust Matter? Antecedents and Contingent Effects of Supervisee Trust on Performance in Selling New Products in China and the United States. Journal of Marketing, Vol. 66, 61-81.

References

Bagozzi, R.P. and Philips, L.W. (1982): Representing and testing Orgnizational Theories: A Holistic Construal. Administrative Science Quarterly, 27, 459-489.

Bansal, P. and Roth, K. (2000): Why Companies go green: A model of ecological responsiveness. Academy of Management Journal, Vol. 43, No 4, 717-736.

Bardhan, P. (2006): Globalization and rural poverty World Development, Vol. 34, No. 8, 1393-1404.

Bardhan, P.K. (1980): Interlocking Factor Markets and Agrarian: A Review of issues. Oxford Economics Papers, New Series Vol.32, No. 1, 82-98.

Basu, K. (2006): Globalization, poverty, and inequality: What is the relationship? What can be done? World Development, Vol. 34, No. 8, 1361-1373, .

Basu, S. (1997): Why Institutional Credit Agencies are Reluctant to Lend to the Rural Poor: A Theoretical Analysis of the Indian Rural Credit Market. World Development, Vol. 25, No. 2, 267-280.

Bhattacharya, C.B. and Sen, S. (2004): Doing Better at Going Good: When, Why and How Consumers Respond to Corporate Social Initiatives: California Management Review, Vo. 47, No 1; 9-24 .

Bhattacharya, C.B., Smith, N.C. and Vogel, D. (2004): Integrating Social Responsibility and Marketing Strategy: An Introduction. California Management Review, Vol. 47, No 1. 6-8.

Bokea, C. and Ikiara, M. (2000): Fishery Commercialization and the Local Economy: The case of Lake Victoria (Kenya): Socio-economics of the Lake Victoria Fisheries. IUCN, Report No.7.

Blair, E and Zinkhan, G.M. (2006): Nonresponse and Generalizability in Academic Research, Editorial, Journal of the Academy of marketing Science, Vol. 34, No. 1, 4-7.

Brashear, T.G., Boles, J.S., Bellenger, D.N and Brooks, C.M (2003): An Empirical Test of Trust-Building Processes and Outcomes in Sales Manager-Salesperson Relationships. Journal of the Academy of Marketing Science. Vol 31, No. 2, 189-200.

Brown, T.J. and Dacin, P.A. (1997): The Company and the Product: Corporate Associations and Consumer Product Reponses, Journal of Marketing, 61/1: 68-84.

Brown, J.R. Dev, C.S. and Lee, D-J (2000): Managing Marketing Channel Opportunism: The Efficacy of Alternative Governance Mechanisms. Journal of Marketing, Vol. 64 51-65 .

Brox O. (1990): The common property theory: epistomological status and analytical utility. Human Organization 49(3): 227 - 235.

Brundtland, G.H. (Ed.) (1987): Our common future; World Commission on Environment and Development. Oxford University Press.

Burnham, T.A., Frels, J.K. and Mahajan, V. (2003): Consumer Switching Costs: A Typology, Antecedents and Consequences. Journal of the Academy of Marketing Science, Vol. 31 No.2, 109-126.

Buvik, A. and John, G. (2000): When does vertical coordination improve industrial purchasing relationships? Journal of Marketing, 64, 52-64.

Byrne, B.M. (1998): Structural Equation Modelling with LISREL, PRELIS and SIMPLIS: Basic Concepts, Applications and programming. Lawrence Erlbaum Associates, 412p, ISBN0-08058-2924-5.

Cannon, J.P., Achrol, R.S. and Gundlach, G.T (2000): Contracts, Norms, and Plural Form Governance. Journal of the Academy of Marketing Science, Vol. 28, No. 2, 180-194.

Cannon, J.P. and Homburg, C. (2001) Buyer-Supplier Relationships and Customer Firm Costs, Journal of Marketing, Vol. 65, 29-43.

Churchill, G.A. (1979): A Paradigm for developing Better Measures of Marketing Constructs, Journal of Marketing Research (February) 64 - 73.

Coase, R.H. (1937): The nature of the firm, Economica, 4, pp. 386-405.

Cohen, J. and Cohen, P. (1983): Applied Multiple Regression/Correlation Analysis for Behavioural Sciences. Hillsdale, NJ: Lawrence Erlbaum and Associates.

Coleman, B.E. (2006): Microfinance in Northeast Thailand: Who Benefits and How Much? World Development, Vol. 34, No. 9, 1612-1638.

Coughlan, A.T. Anderson, E, Stern, L.W. El-Ansary, A.I (2001): Marketing Channels, Sixth Edition, Prentice Hall, Upper Saddle River, New Jersey.

Delgado, C.L., Wada, N., Rosegrant, M.W., Meijer, S. and Ahmed, M. (2003): Fish to 2020, Supply and Demand in Changing Global Markets. IFPRI -WorldFish Center Technical Report 62.

Demsetz, H. (1967): Toward a Theory of Property Rights. American Economic Review, 57: 347-359.

Demsetz, H. (1966): Some aspects of property rights. Journal of Law and Economics 9: 61-70.

Demsetz, H. (1964): The exchange and enforcement of property rights. Journal of Law and Economics 3: 11-26.

Dorward, A. (2001): The Effects of Transaction Costs, Power and Risk on Contractual Arrangements: A Conceptual Framework for Quantitative Analysis. Journal of Agricultural Economics - Volume 52, Number 2, pp 59-73.

Eisenhardt, K.M. (1989a): Building Theories from Case Study Research. Academy of Management Review, Vol. 14, No. 4, 532-550.

Eisenhardt K.M. (1989b). Agency theory: an assessment and review. Academy of Management Review 14: 57-74.

Ellis, F. (1988): Peasant economics, farm household and agrarian development, Cambridge University Press, Cambridge UK.

Fafchamps, M. (2004): Market Institutions in Sub-Saharan Africa. Theory and Evidence. The MIT Press, Cambridge, Massachusetts, London, England.

Fafchamps, M. (1996): The Enforcement of Commercial Contracts in Ghana World Development, Vol. 24, No. 3, pp. 427448. 1996.

FAO (2001): Contract farming, Partnership for growth. FAO Agricultural Services, Bulletin 145. Food and Agriculture Organization of the United Nations.

FAO (1995) Quality and changes in fresh fish: FAO Fisheries Technical Papers - T348, 150pp.

Feld, S.L. (1981): The Focussed Organization of Social Ties. The American Journal of Sociology, Vol. 86, No.5 (March, 1981) 1015-1035.

Fornell, C. and Larcker, D.F. (1981): Evaluating Structural Equation Models with Unobservable Variables and Measurement Error. Journal of Marketing Research, Vol. XVIII, 39-50.

Foss, K. and Foss, N.J. (2005): Resources and Transaction Costs: How Property Rights Economics furthers the Resource-Based View. Strategic Management Journal, 26: 541-553 .

Ganesan, S and Hess, R. (1997): Dimensions and Levels of Trust: Implications for Commitment to a Relationship. Marketing Letters 8:4 (1997): 439-448.

Ganesan, S. (1994), "Determinants of Long-Term Orientation in Buyer-Seller Relationships," Journal of Marketing, 58 (2), 1-19.

References

Geheb, K. (ed.) (2002). Report of the LVFRP Nutrition Survey LVFRP Technical Document No. 18. LVFRP/TECH/00/18. The Socio-economic Data Working Group of the Lake Victoria Fisheries Research Project, Jinja.

Geheb, K. (1997). The regulators and the regulated: fisheries management, options and dynamics in Kenya's Lake Victoria Fishery. Unpublished D.Phil.Thesis, University of Sussex, Falmer, Brighton, U.K. Reprinted as LVFRP Technical Document No.10 LVFRP/TECH/00/10. The Socio-economic Data Working Group of the Lake Victoria Fisheries Research Project, Jinja.

Geyskens, I., Steenkamp J.B.E.M. and Kumar, N. (2006): Make, buy or Ally: A Transaction Cost theory Meta-analysis: Academy of Management Journal, Vol. 49, No.3 519-543).

Geyskens, I., Steenkamp J.B.E.M. and Kumar, N. (1998): Generalizations about trust in marketing channel relationships using meta analysis. International Journal of Research in Marketing, 15 (1998) 223-248 .

Ghosh, M. and John, G. (1999): "Governance Value Analysis and Marketing Strategy," Journal of Marketing, 63 (Special Issue), 131-45.

Gibbon, P. (1997): Of Saviors and Punks: The political economy of the Nile perch marketing chain in Tanzania. CDR Working Paper 97 (www.cdr.dk) .

Glover, D.J. (1987): Increasing the Benefits to Smallholders from Contract Farming: Problems for Farmers' Organizations and Policy Makers. World Development, Vol. 15, No. 4, 441-448.

Glover, D.J. (1984): Contract Farming and Smallholder Outgrower Schemes in Less-developed Countries World Development, Vol. 12, Nos. 11/12, 1143-1157, 1984.

Granovetter, M.S. (1985): Economic Action and Social Structure: The Problem of Embeddedness. American Journal of Sociology, 91 (3), 481-510.

Granovetter, M.S. (1973): The Strength of Weak Ties: American Journal of Sociology, Vol. 78, No. 6 (May 1973) 1360-1380.

Green, P.E. and Srinivasan, V. (1990): Conjoint analysis in marketing: New developments with implications for research and practice. Journal of Marketing 54 (October), 3-19.

Green, P.E. and Srinivasan, V. (1978). Conjoint Analysis in Consumer Research. Issues and outlook. Journal of Consumer Research. Vol. 5, September 1978.

Green, P.E., Krieger, A.M. and Wind, Y.J. (2001): Thirty Years of Conjoint Analysis: Reflections and Prospects. INTERFACES 31: 3, Part 2 of 2, May-June 2001 (pp. S56-S73).

Grewal, R. and Dharwadkar, R (2002): The Role of the Institutional Environment in Marketing Channels. Journal of Marketing, Vol. 66, 82-97.

Grosh, B (1994): Contract Farming in Africa: An Application of the New Institutional Economics, University, Journal of the African Economies, Volume 3, Number 2 pp231-261.

Gundlach, G.T. and Cadotte, E.R. (1994): Exchange Interdependence and Inter-firm Interaction: Research in a Simulated Channel Setting," Journal of Marketing Research, 31 (November), 516-32.

Hair, J.F., Anderson, R.E., Tatham, R.L. and Black, W.C. (1998): Multivariate Data Analysis. Fifth Edition, Prentice Hall International INC.

Harris-White, B. eds. (1999): Agricultural Markets from Theory to Practice. Field experiences in Developing Countries. Basingstoke, Macmillan Press.

Hardin, G. (1968): The Tragedy of the Commons. Science 162, 1243-1248.

Heide, J.B. (2003): Plural governance in Industrial Purchasing. Journal of marketing, Vol. 67 (October), 18-29.

Heide, J.B. (1994). Inter-organizational governance in marketing channels: Theoretical perspectives on forms and antecedents. Journal of Marketing, 58, 71-85.

Heide, J.B. and John, G. (1992): Do norms matter in marketing relationships? Journal of Marketing, 56, 32-44.

Heide, J.B. and John, G. (1988): The Role of Dependency Balancing in Safeguarding Transaction - Specific Assets in Conventional Channels. Journal of Marketing, 52, 20-35.

Henderson, J.M. and Quandt, R.E (1985). Micro-economic Theory. A Mathematical Approach. Third Edition, McGraw-Hill, Inc.

Hendrikse, G. (2004): Governance in Chains and Networks. In The Emerging World of Chains and Networks: Bridging Theory and Practice: Camps, T, Diederen, Hofstede, G.J and Vos, B. ed: Reeds Business Information (p189- 204).

Henson, S., Brouder, A.M. and Mitullah, W (2000): Food Safety Requirements and Food Exports from Developing Countries: The Case of Fish Exports from Kenya to the European Union: American Journal of Agriculture Economics: 82(5), 1159-1169.

Henson, S. and Loader, R. (2001): Barriers to Agricultural Exports from Developing Countries: The Role of Sanitary and Phytosanitary Requirements. World Development, (29) 1: 85-102.

Henson, S. and Traill, B. (1993): The demand for food safety: market imperfections and the role of government. Food Policy (18): 152-162.

Henson, S. and Mitullah, W. (2003): Kenyan Exports of Nile perch: Impact of Food safety standards on an Export-Oriented Supply chain. University of Guelph and University of Nairobi.

Hewett, K. and Bearden, W.O. (2001): Dependence, Trust, and Relational Behavior on the Part of Foreign Subsidiary Marketing Operations: Implications for Managing Global Marketing Operations. Journal of Marketing Vol. 65 (October 2001), 51-66.

Hibbard, J.D, Kumar, N and Stern, L.W. (2001): Examining the Impact of Destructive Acts in Marketing Channel Relationships. Journal of Marketing Research Vol. XXXVIII (February 2001), 45-61 45.

Hill, C.W. L. and Jones, T.M. (1992): Stakeholder-Agency Theory. Journal of Management Studies 29 (2): 131-154.

Holling, C.S. (2000): Theories for Sustainable Futures. Conservation Ecology 4(2): 7. www. ecologyandsociety.org.

Houston, M.B. and Johnson, S. A. (2000): Buyer-Supplier Contracts Versus Joint Ventures: Determinants and Consequences of Transaction Structure. Journal of Marketing Research Vol. XXXVII (February 2000), 1-15 .

Hunt S.D. and Lambe C.J. (2000): Marketing's Contribution to business strategy: Market orientation, relationship marketing and resource-advantage theory. International Journal of Management Reviews Volume 2 (i) pp17-43.

Ingenbleek, P. and Meulenberg, M.T.G. (2006): The Battle Between "Good" and "Better": A Strategic Marketing Perspective on Codes of Conduct for Sustainable Agriculture, Agribusiness, Vol. 22 (4) 451-473.

References

Jap S.D. and Ganesan, S. (2000): Control Mechanisms and the Relationship Life Cycle: Implications for Safeguarding Specific Investments and Developing Commitment. Journal of Marketing Research Vol. XXXVII (May 2000), 227-245 227.

Jansen, E.G. (1997): Rich Fisheries - Poor fishfolk. Some preliminary Observations about the effects of Trade and Aid in the Lake Victoria Fisheries. Socio-economics of the Lake Victoria Fisheries. IUCN, Report No. 1 September 1997.

Jayne, T.S., Govereh, J., Mwanaumo, A., Nyoro, J.K. and Chapoto, A. (2002): False Promise or False Premise? The Experience of Food and Input Market Reform in Eastern and Southern Africa. World Development Vol. 30, No. 11, 1967-1985, 2002.

Jöreskog, K.G. and Sörbom, D. (2005): LISREL 8.7 for Windows [Computer Software]. Lincolnwood, IL: Scientific Software International, Inc.

Joshi, A.W. and Campbell, A.J (2003): Effect of Environmental Dynamism on Relational Governance in Manufacturer- Supplier Relationships: A Contingency Framework and an Empirical Test. Journal of the Academy of Marketing Science, Vol. 31, No. 2, 176-188.

Joshi, A.W. and Stump, R.L (1999): The Contingent Effect of Specific Asset Investments on Joint Action in Manufacturer-Supplier Relationships: An Empirical Test of the Moderating Role of Reciprocal Asset Investments, Uncertainty, and Trust. Journal of the Academy of Marketing Science, Vol. 27, No. 3, pages 291-305.

Jul-Larsen, E., Kolding, J., Overa, R., Nielsen, J.R. and Zwieten, P.A.M. van (2003): Management, co-management or no management? Major dilemmas in Southern African Freshwater Fisheries. Part 1: Synthesis report. FAO Fisheries Technical paper No 426/1. Rome, FAO. 2003. 126p.

Kambewa, E., Ingenbleek, P. and van Tilburg, A. (2006): Stretching Corporate Social Responsibility Upstream: Improving Sustainability with Upstream Partners in Global Marketing Channels: Paper presented at the conference on Corporate Responsibility and Global Business: Implications for Corporate and Marketing Strategy, London Business School, United Kingdom, July, 13-14, 2006.

Kaplinsky, R. and Morris, M. (1999): Trade Policy Reform and the Competitive Response in Kwazulu Natal Province, South Africa. World Development Vol. 27, No. 4, 717-737, 1999.

Kaul, A. and Rao, V.R. (1995): Research for product positioning and design decisions: An integrative review " International Journal of Research in Marketing, 12, 293-320.

Kenya Government (1991): The Laws of Kenya: The Fisheries Act. CAP 378 (Revised 1991). Government Printer, Nairobi. Pp1-90.

Key, N. and Runsten, D. (1999): Contract Farming, Smallholders, and Rural Development in Latin America: The Organization of Agroprocessing Firms and the Scale of Outgrower Production. World Development Vol. 27, No. 2, pp. 381-401, 1999.

Kim, J. and Mahoney, T.T. (2005): Property rights theory, transaction costs theory, and agency theory: An Organizational Economics Approach to Strategic Management, Managerial and Decision Economics, Vol.26, Issue 4, 223-242.

Kim, J. and Mahoney, T.T. (2002): Resource-based and property rights perspectives on value creation: The case of oil field unitization. Managerial and Decision Economics, Vol. 23, Issue 4-5, p225-245.

Kim, K. (1999): On determinants of joint action in industrial distributor-supplier relationships: Beyond economic efficiency: International Journal of Research in Marketing 16 (1999) 217-236.

Kirmani, A. and Rao, A.R. (2000): "No Pain, No Gain: A Critical Review of the Literature on Signaling Unobservable Product Quality," *Journal of Marketing*, 64 (2), 66-79.

Kish, L. (1965). Survey sampling. New York: Wiley.

Lake Victoria Fisheries Organization (LVFO) (2000): Summary of the Lake Victoria fisheries frame survey, March 2000.

Lake Victoria Fisheries Organization (LVFO) (1999): Strategic Vision for Lake Victoria (1999-2015) Jinja, Uganda.

Langerak, F. (2001): Effects of Market Orientation on the behaviours of salespersons and purchasers, channels relationships, and performance of manufacturers. International Journal of Research in marketing 18 (2001) 221-234.

Larson A. (1992). Network dyads in entrepreneurial settings: a study of governance of exchange relationships. Administrative Science Quarterly 37: 76-104.

Leeflang, P.S.H. and van Raaij, W.F. (1995): The changing consumer in the European Union: A "meta-analysis". Intern. J. of Research in Marketing 12 (1995) 373-387.

Luo, Y. (2002): Contract, Cooperation, and Performance in International Joint Venture. Strategic Management Journal 23 (2002) 903-919.

Luning, P.A., Marcelis, W.J. and Jongen, W.M.F. (2002): Food quality management - a techno-managerial approach. Wageningen, The Netherlands, Wageningen Pers.

Lusch, R.F. and Brown, J.R. (1996): Interdependence, Contracting and Relational Behavior in Marketing Channels. Journal of Marketing 60 (October): 19-38.

Luten, J.B., Oehlensechläger, J. and Ólafsdóttir, G. (eds) (2003): Quality of Fish from Catch to Consumer, Labelling, Monitoring and Traceability: Wageningen Academic Publishers. pp 456.

MacKenzie, S.B., Podsakoff, P.M. and Paine, J.B. (1999): Do Citizenship Behaviours Matter More for Managers then for Salespeople. Journal of Academy of Marketing Science, Vol. 27, No.4, 396-410.

Maddala, G.S. (1992): Introduction to Econometrics. Second Edition, MacMillan, New York .

Mahoney, J.T. (2001): A resource-based theory of sustainable rents. Journal of Management 27, 651-660.

Maignan, I. and Ferrell, O.C. (2004): Corporate Social Responsibility and Marketing: An Integrative Framework, Journal of the Academy of Marketing Science. Vol. 32, No. 1, 3-19.

Maignan, I. and Ferrell, O.C. (2003): Nature of corporate responsibilities: Perspectives from American, French, and German consumers. Journal of Business Research, 56, 55- 67.

Maignan, I., Ferrell, O.C. and Hult, T.M. (1999): Corporate Citizenship: Cultural Antecedents and Business Benefits, Journal of the Academy of Marketing Science, 27(4), 455-469.

Malhotra, D. and Murnighan, J.K. (2002): The effect of contracts on interpersonal trust, Administrative Science Quarterly, Vol. 47, No 3, 534-559.

Masakure, O. and Henson, S. (2005): Why do small-scale producers choose to produce under contract? Lessons from Nontraditional Vegetable Exports from Zimbabwe, World Development, Vol, 33, Issue 10, October 2005, 1721-1733 .

Maslow, A. (1943): A theory of human motivation, Psychological Review, vol. 50, 1943, 370-96.

Moorman, C., Zaltman, G. and Deshpande, R. (1992): Relationships between providers and users of market research: The dynamics of trust within and between organizations. Journal of Marketing Research, 29, 314-328.

Morgan, R.M. and Hunt, S.D. (1994), "The Commitment- Trust Theory of Relationship Marketing," Journal of Marketing, 58, 20-38.

Ng, F. and Yeats, A. (1997): Open Economies Work Better! Did Africa's Protectionist Policies Cause its Marginalization in World Trade? World Development, Vol. 25, No. 6, 889-904, .

Nissanke, M. and Thorbecke, E. (2006): Channels and policy debate in the globalization-inequality-poverty nexus World Development, Vol. 34, No. 8, 1338-1360, .

North, D. (1990): Institutions, institutional change and economic performance. Cambridge University Press, Cambridge.

Olafsdottir, G., Nesvadba, P., Natale, C.D., Careche, M., Oehlenschlager, J., Tryggvadottir, S.V., Schubring, R., Kroeger, M., Heia, K., Esaiassen, M., Macagnano, A. and Jorgensen, B.M. (2004): Multisensor for fish quality determination. Trends in Food Science and Technology, 15, 86-93.

Otsuki, T., Wilson, J.S. and Sewadeh, M. (2001): Saving two in a billion: quantifying the trade effect of European food safety standards on African exports. Food Policy 26, 495-514.

Owino, J.P. (1999): Traditional and Central management systems of the Lake Victoria Fisheries in Kenya: Socio-economics of the Lake Victoria, IUCN. Report No 4.

Pfeffer, J. (1982): Organizations and Organizational Theory. Marshfield, MA: Pitman.

Pfeffer and G.R. Salancik (1978): The External Control of Organizations. New York: Harper and Row.

Pindyck, R.S. and Rubinfeld, D.L (1998): Econometric Models and Economic Forecasts. McGraw-Hill International Editions, 4th Edition.

Ping, R.A. (1997): Voice in Business - to - Business Relationships: Cost-of-exit and Demographic Antecedents. Journal of Retailing, Vol. 73 (2), 261-281.

Platteau, J-P. and Abraham, A. (1987): An inquiry into quasi-credit contracts: The role of reciprocal credit and interlinked deals in small-scale fishing communities. Journal of development studies 23 (4), 461-490.

Poate, C.D. and Daplyn, P.F. (1993): Data for Agrarian Development. Cambridge University. Cambridge University Press.

Porter, G. and Phillips -Howard, K. (1997): Comparing Contracts: An Evaluation of Contract Farming Schemes in Africa. World Development, Vol. 25, No. 2, 227-238.

Porter, M.E. (1980): Competitive Strategy. Techniques for analyzing industry and competitors: New York: The Free Press.

Porter, M.E. and Kramer, M.E. (2006): Strategy and Society. The Link between Competitive Advantage and Corporate Social Responsibility. Harvard Business Review, December (2006).

Poppo, L. and Zenger, T. (2002): Do Formal Contracts and Relational Governance Function as Substitutes or Complements? Strategic Management Journal, 23, 7007-725.

Prahalad, C.K. and Hammond, A. (2002): Serving the Worlds' Poor Profitably. Harvard Business Review, September.

Prahalad, C.K. and Hart, S.L. (2002): The Fortune at the Bottom of the pyramid. Strategy and Business, 26, 54-67.

Rahman, A. (1999): Micro-credit Initiatives for Equitable and Sustainable Development: Who Pays? World Development, Vol. 27, No. 1, 67-82, 1999.

Ramamurti, R. (1999): Why Haven't Developing Countries Privatized Deeper and Faster? World Development, Vol. 27, No. 1, 137-155, 1999.

Ravallion, M. (2006): Looking beyond averages in the trade and poverty debate: World Development Vol. 34, No. 8, 1374-1392 .

Rigg, J. (2006): Land, farming, livelihoods, and poverty: Rethinking the links in the Rural South, World Development, Vol. 34, Issue 1, 180-202.

Rigg, J. (1998): Tracking the poor: The making of wealth and poverty in Thailand (1982-1994). International Journal of Social Economics, 25(6-8), 1128-1141.

Rindfleisch, A. and Heide, J.B. (1997): Transaction Cost Analysis: Past, Present, and Future Applications," Journal of Marketing, 61, 30-54.

Rindfleisch, A. and Moorman, C. (2001): The Acquisition and Utilization of Information in New Product Alliances: A Strength-of-Ties Perspective, Journal of Marketing, 65, 1-18.

Roex, J. and Miele, M., eds. (2005): Farm Animal Welfare Concerns: Consumers, Retailers and Producers. Welfare Quality; Science and Society improving animal welfare. Welfare Quality Report No 1.

Rokkan, A.I. Heide, J.B. and Wathne, K.H. (2003): Specific Investments in Marketing Relationships: Expropriation and Bonding Effects: Journal of Marketing Research Vol. XL (May 2003), 210-224.

Saenz-Segura, F. (2006): Contract Farming in Costa Rica: Opportunities for smallholders? PhD thesis, Wageningen University.

Sauper, H. (2004): The Darwin's Nightmare, A documentary film: www.darwinsnightmare.com.

Sen, S., Bhattacharya, C.B. and Korschun, D. (2006): The Role of Corporate Social Responsibility in Strengthening Multiple Stakeholder Relationships: A Field Experiment Journal of the Academy of Marketing Science. Vol 34, No. 2, 158-166.

Sen, S. and Bhattacharya, C.B. (2001): Does Doing Good Always Lead to Doing Better? Consumer Reactions to Corporate Social Responsibility. Journal of Marketing Research, 225 Vol. XXXVIII, 225-243.

Scherer, F.M. and Ross, D. (1990): Industrial Market Structure and Economic Performance. Third Edition, Houghton Mifflin Company, Boston .

Schermelleh-Engel, K., Moosbrugger, H. and Müller, H. (2003): Evaluating the Fit of Structural Equation Models: Tests of Significance and Descriptive Goodness-of-Fit Measures, Methods of Psychological Research, Vol.8, No.2, 23-74.

Sirdeshmukh, D., Singh, J. and Sabol, B. (2002): Consumer Trust, Value, and Loyalty in Relational Exchanges. Journal of Marketing, Vol. 66, 15-37.

Shook, C.L., Ketchen, D.J. Jr, Hult, G.T.M. and Kacmar, K.M. (2004): An Assessment of the use of Structural Equation Modelling in Strategic Management Research, Research notes and Commentaries. Strategic management Journal, 25: 397-404.

Singh, S. (2002): Contracting Out Solutions: Political Economy of Contract Farming in the Indian Punjab. World Development, Vol. 30, No. 9, 1621-1638, 2002.

Smith, N.C. (2003): Corporate Social Responsibility: Whether or How? California Management Review, 45 (4), 52-76.

Southgate, D., Salazar-Canelos, P., Camacho-Saa, C. and Stewart, R. (2000): Markets, Institutions, and Forestry: The Consequences of Timber Trade Liberalization in Ecuador. World Development, Vol. 28, No. 11, 2005-2012.

SPSS Inc. (2003): SPSS® Base 12.0 User's Guide: 233 South Wacker Drive, 11th Floor, Chicago, IL 60606-6412.

Stern, L.W., El-Ansary, A.I and Coughlan, A. (1996): Marketing Channels. Englewood Cliffs, NJ: Prentice Hall.

Stewart, D.W. and Shamdasani, P.N. (1990): Focus Group: Theory and Practice. Applied social research methods series, Vol. 20, ISBN 0-0839-3389-4; 153p.

Suchman, M.C. (1995): Managing Legitimacy: Strategic and Institutional approaches. Academy of Management Review, 20: 571-610.

Sudman, S. (1976): Applied sampling. New York: Academic Press.

The Economist (2001a): "Mobile phones in India, Another kind of network" March 1, 2001.

The Economist (2001b): "Fishermen on the net," November 8, 2001.

The Economist (2005a): "The real digital divide" March 10, 2005.

The Economist (2005b): "Calling across the divide" March 10, 2005.

Thorbecke, E. and Nissanke, M. (2006): The impact of globalization on the world's poor, Introduction: World Development, Vol. 34, No. 8, pp. 1333-1337.

Uzzi, B. (1997): Social Structure and Competition in Interfirm Networks: The Paradox of Embeddedness, Administrative Science Quarterly, 42 (1), 35-67.

Vachani, S. and Smith, N.C. (2006): Socially Responsible Distribution: Distribution Strategies for Reaching the Bottom of the Pyramid. Paper presented at the Conference on "Corporate Responsibility and Global Business: Implications for Corporate and market Strategy."London Business School, July 2006.

Vachani, S. and Smith, N.C. (2004): Socially Responsible Pricing: Lessons from the Pricing of Aids Drugs in Developing Countries. California Management Review, Vol. 47, No 1. 117-144 .

Van der Hoff, F. and Roozen, N (2001): Fair Trade: The Story behind Max Harvelaar-Coffee, Oké-bananas and Kuyichi-Jeans. Van Gennep (in Dutch).

Van der Laan, H.L., Dijkstra T. and Van Tilburg, A., Eds. (2000): Agricultural Marketing in Tropical Africa. Aldershot, Ashgate.

Van der Meer, C.I.J. (2006): Exclusion of Small-scale Farmers from coordinated supply chains Market failure, policy failure or just economies of scale? In: Ruben, Slingerland, and Nijhoff (eds): Agro-Food Chains and Networks for Development (Chapter 18). Frontis/Springer.

Van Tilburg, A., Moll H.A.J. and Kuyvenhoven, A. Eds. (2000): Agricultural markets beyond Liberalization. Boston/London Kluwer Academic Publishers.

Verhallen, T., Wiegerinck, V., Gaaker, C. and Poiesz, T. (2004): Demand Driven Chains and Networks. In The Emerging World of Chains and Networks: Bridging Theory and Practice: Camps, T, Diederen, P., Hofstede, G.J and Vos, B. eds: Reeds Business Information (129- 146).

Wade, R.H. (2004): Is Globalization Reducing Poverty and Inequality? World Development, Vol. 32, No. 4, 567-589.

Wathne, K.H. and Heide J.B. (2004): Relationship Governance in a Supply Chain Network Journal of Marketing Vol. 68, 73-89.

Wathne, K.H. Biong, H. and Heide, J.B. (2001): Choice of Supplier in Embedded Markets: Relationship and Marketing Program Effects. Journal of Marketing Vol. 65, 54-66.

Wathne, K.H. and Heide, J.B. (2000): Opportunism in Interfirm Relationships: Forms, Outcomes, and Solutions. Journal of Marketing Vol. 64, 36-51.

Wilkie, W.L. and Moore, E.S. (1999): Marketing's contributions to Society: Journal of Marketing, Vol.63 (Special issue), 198 -218.

Williamson, O.E. (1985): The Economic Institutions of Capitalism. The Free Press, MacMillian, Inc., New York.

Wittink, D.R., Vriens, M and Burhenne, W. (1994): Commercial use of conjoint analysis in Europe: Results and critical reflections, International Journal of Research in Marketing 11, 41-52.

World Bank (2003): World development report 2003; Sustainable Development in a Dynamic World. Transforming Institutions, Growth and Quality of Life. Washington, DC: World Bank.

Wrong, D.H. (1968): Some problems of Defining Social Power. The American Journal of Sociology, Vol. 73, No. 6, 673-681.

Wuyts, S. and Geyskens, I. (2005): The Formation of Buyer-Supplier Relationships: Detailed Contract Drafting and Close Partner Selection, Journal of Marketing, Vol. 69, 103-117.

Wuyts, S, Stremersch, S., van den Bulte, C. and Franses, P.H. (2004): Vertical Marketing Systems for Complex Products: A Triadic Perspective. Journal of Marketing Research, Vol. XLI, 479-487.

Yin, R.K. (2003): Case Study Research Design and Methods. Third edition. Applied Social Research Methods Series, Vol. 5. Sage Publications.

Zeithaml, V.A. (1988): Consumer Perceptions of Price, Quality and Value: A Means -End Model and Synthesis of Evidence. Journal of Marketing, Vol. 52, 2-22.

Summary

Sustainable development hinges on a combined focus of its impact on society (people), the environment (planet) and to its economic (profit) value (Brundlandt, 1987). Increasingly, it is being recognised that the people, profit and planet (PPP) dimensions are interlinked and an important challenge for public and private policy is to take them jointly into account. This interlinkage is particularly evident in international chains that build on scarce natural resources from developing countries, which is the key focus of this thesis. Although globalization and market integration promise opportunities for economic development, poverty and food insecurity continue to be widespread in the less developed segments of the global society. Due to limited possibility to switch to other sources of livelihood, poor communities in developing economies are relying on increasingly scarce natural resources. Yet, small-scale primary producers in developing economies continue to be marginalized or excluded from the global networks thereby limiting the chances of benefiting from the opportunities that may come along with increased market integration.

This thesis investigates how international marketing channels can be organized in order to better balance the PPP dimensions such that small-scale primary producers from developing economies are integrated into the international marketing channels in a way that adds to the profitability of the chain, the welfare of the local communities, without compromising the sustainability of natural resources. The central question of the study is how channel members collectively can take joint responsibility and action for a good balance of the PPP issues in the chain. The study focuses on whether contracts might be used to establish a better PPP balance in international marketing channels and what should be the terms of contract and governance structures that may stimulate welfare, profitability and sustainability at the primary production level and in the channel as a whole. The thesis addresses the following questions, (1) What market failures constrain the ability of primary producers to establish a better balance of PPP in their activities? (2), What terms of contracts and governance structures can stimulate a better balance of welfare, sustainability and quality at primary production? and (3), In what way can downstream channel partners and other stakeholders help to stimulate welfare, sustainability and quality in the channel and especially at the primary production and what would motivate them to do so?

In order to answer these questions, Chapter 2 sets out a theoretical framework that reviews the different theories of channel governance highlighting their limitations in promoting sustainable practices. Consequently, the chapter integrates a number of theoretical perspectives, i.e., the transaction costs economic, social theory, network theory and property rights theory to develop arguments for the use of the contracts to promote sustainable and quality-enhancing practices at primary production level that are tested in subsequent chapters. Chapter 2 argues that the choice of contracts depends on the market failures to be solved and the environment in which contracts are applied such as the social environment, profitable markets, physical environment, regulatory environment and property rights.

The theoretical framework is operationalized in a research setting introduced in Chapter 3. It undertakes a situational analysis of the primary stages of the Nile perch channel in Kenya to identify the exact market failures that primary producers face. An analysis of these market failures provides background insights for the understanding of the terms of contract and governance structures that may stimulate sustainable and quality-enhancing practices at primary production level. Primary producers face complex and intertwined market failures such as declining fisheries production, limited access to production facilities, information asymmetries and ineffective enforcement of sustainable practices.

The degradation of the fisheries means that fishermen catch less fish for their business interests as well as own welfare needs. In absence of alternative sources of livelihoods to relieve the fisheries from pressure exerted by both welfare demands and business interests, fishermen tend to be compelled to use bad fishing gears so as to enable them to catch at least some fish. In that regard, the use of bad fishing gears is both a result and a cause for the degradation of the fisheries. The use of bad fishing gears is also attributed to high prices of the recommended fishing gears and ineffective enforcement for sustainable practices by relevant authorities. Fishermen argue that recommended fishing gears are relatively more expensive compared to bad gears which are also readily available. Although some fishermen can afford to buy the good fishing gears, they may be opportunistic because they would not be easily detected anyway due to ineffective enforcement. In turn, fishermen who may be willing to undertake sustainable fishing practices may be less motivated to do so partly due to the fact that a level playing field is not being guaranteed.

Inevitably, alternative sources of livelihoods may not become available overnight. Therefore, improving the income from the limited fish production and minimizing the risks that fishermen face in their daily transactions might be an option in order to stimulate sustainable and quality-enhancing practices in an already difficult environment. The potential benefits that primary producers would realize from the limited production are often compromised by lack of production facilities (fishing gears for some and tools for quality improvement for almost all) and information asymmetries that compromise their ability to bargain for competitive terms of transactions from buyers who tend to own production facilities and also have market information. Although some of the market failures that fishermen face such as information asymmetries, provision of production facilities and enforcement could potentially be addressed through contracts, the question is whether contracts would motivate fishermen to adopt sustainability and quality-enhancing practices and if so, what should be the terms of contract.

In order to answer this question, Chapters 4 and 5 design contracts that offer possible solutions to information asymmetry, limited access to production facilities, ineffective enforcement and better integration of fishermen into international channels. Most importantly, the contracts oblige fishermen (Chapter 4) to implement and middlemen (Chapter 5) to drive sustainable and quality-enhancing practices. In both chapters, preference for sustainability and quality-enhancing contracts is expected to be influenced by the terms of contracts in interaction with the context in which fishermen and middlemen operate.

The results of Chapter 4 show strong support for our conceptual framework that fishermen's preference for sustainability and quality-enhancing contracts depends on the terms of contract in interaction with contextual factors that they face. We find strong fishermen's preference for sustainability and quality-enhancing contracts especially those that provide production facilities, price information, that bring fishermen closer to international channels and that allow private policy enforcement of sustainable practices. Interestingly also, the analyses show that increasing decline of fisheries production as evidenced by increasing catch uncertainty increases fishermen's preference for contracts in which recommended fishing gears are provided to promote sustainability and, particularly fishermen that incur more quality losses prefer to have recommended fishing gears for improving sustainability instead of quality-improving tools for short-term economic gains. However, there is no single contract that would suit the needs of all; fishermen differ in their preferences for specific terms of contracts. Considering the primary motivation for contracts, about 24% of fishermen prefer contracts in which price information is provided and about 47% of fishermen prefer contracts in which recommended fishing gears are provided. About 10% of fishermen prefer contracts in which tools for quality improvement are provided. About 13% of fishermen were almost indifferent between contracts with processing factories and also contracts in which fish prices are fluctuating. The remaining 6% primarily preferred contracts with middlemen. These results suggest that no single contract can be applied and therefore, different contracts may have to be designed and different fishermen would select the ones that best fit their preferences and circumstances

The results in Chapter 5 also support our conceptual framework in that middlemen's preference for contracts is influenced by the terms of contract in interaction with contextual factors that they face. At their purchase side, we find that middlemen prefer short-term contracts with fishermen in which price information is provided. Preference for the terms of contracts is moderated by middlemen's social relations with fishermen. We find that middlemen that have conflicts with fishermen prefer contracts with new fishermen while those that trust fishermen prefer to contract the same fishermen and are also willing to give price information to fishermen. Middlemen that have high level of dependence on fishermen prefer contracts with fishermen in which price information is provided. Similarly, middlemen that give fishing gears to fishermen prefer contracts with current fishermen and also contracts in which price information is provided. We also find that middlemen that currently face quality losses have a preference for contracts with new fishermen and those that face supply uncertainty prefer short-term contracts. At their supply side, middlemen prefer long-term contracts with processing factories moderated by the number of network ties such that middlemen that have many buyers prefer short term contracts. We find that network ties also moderate middlemen's preference for fluctuating factory prices with a stronger preference for fluctuating prices for middlemen that have many buyers.

These results, especially at the purchase side of the middlemen, show that there is room for improvement in the provision of price information which would suit the needs and preference of both the fishermen and the middlemen. Such contract terms would minimise the conflicts between fishermen and middlemen that often arise over unpredictable price fluctuations. At the supply side, we find no effect of social relations on middlemen's preference for the terms of contracts with processors. With this lack of evidence, we raised a question about whether social relations may influence the choice of governance structure in an oligopsony and also where suppliers operate at much smaller scale than buyers as was the case in middlemen/processors transactions.

In Chapter 6, we take the analysis across the supply chain one step further in addressing the question: in what way can downstream channel partners and other stakeholders help to stimulate welfare, sustainability and quality-enhancing practices in the channel and especially at the primary production level and what would motivate them to do so? The chapter builds on corporate social responsibility literature to examine how and why downstream members may take responsibility and actions to solve the upstream PPP problems. The chapter extends the analysis from a single company-society context to a channel–society context where the success of a channel member's responsibility and actions in balancing the PPP issues may influence or be influenced by those of members in other stages of the channel. In order to address this question, the chapter investigates the extent to which downstream channel members are aware of upstream problems, take responsibility over the problems and are willing to take action to solve the problems. The results show a diminishing awareness about upstream problems as the channel goes farther downstream, channel members take partial responsibility over the PPP problems and show conditional willingness to take action to solve the problems.

The diminishing awareness about upstream problems farther downstream is attributed to the information asymmetries created in intermediary stages of the channel where hiding or misrepresenting the fact about the problems may be out of self-interest. Additionally, the lack of information flow across the whole channel is also attributed to downstream channel members not being very aggressive and proactive in pursuing the knowledge about where the fish comes from and under what conditions it is produced.

In the overall, the chapter shows that as the channel moves farther downstream, not only does awareness of PPP problems diminish, but dependence on the Nile perch specific resources also declines. After all, retailers and importers have a wider global sea in which to cast their nets to meet their supply needs and Nile perch is a substitutable resource without a strong image at the retail outlet. Therefore, resource dependency will not be a motivation in the downstream to help solve the Nile perch channel-specific PPP problems. Their stake is more likely to lie in legitimacy arguments as they arise from stakeholder pressure. Specifically, the channel members may respond to stakeholder pressure to protect other resources (than physical fish) such as their image and identity that may be damaged in the event of negative publicity by stakeholders such as the special interest groups. The chapter also shows that the position and the power of the members in the channel may be an important factor to influence change in favour of the PPP balance in the channel. Namely, the retailers may be in a better position and also have the market power to demand sustainable products from their suppliers and that may likely lead to changes in upstream channel.

In Chapter 7, we present general conclusions, discussion and implications of the results presented in the thesis. We acknowledge that this study is among the first to consider sustainable development in a channel as a whole from primary production to retail distribution. This study also integrates a number of theoretical arguments to argue for the use of contracts in promoting sustainable and quality-enhancing practices. The chapter reflects on the main conclusions from the study namely that fishermen are open to contracts as mechanisms to stimulate sustainable and quality-enhancing practices and that this can only materialise if they are supported by middlemen and downstream channel members and other stakeholders. Middlemen and downstream channel members and other stakeholders could contribute to sustainable development in supporting the implementation of sustainable and quality-enhancing practices by providing price information and production facilities. The chapter also highlights that in order to get downstream channel members involved in balancing the PPP dimensions, the special interest groups could play an important role in putting sustained pressure on the downstream channel members.

Curriculum Vitae

Emma Kambewa, PhD candidate in the Marketing and Consumer Behaviour Group of Wageningen University, obtained her BSc in Agriculture (1992) and her MSc degree in Agricultural Economics from the University of Malawi (1998). She joined Wageningen University in 2003 under a multidisciplinary research program on governance and quality in perishable product chains. The title of her PhD thesis (2007) is: "Balancing the People, Profit and Planet Dimensions in International Marketing Chains: A study on coordinating mechanisms in the Nile perch channel from Lake Victoria". It focuses on the question how international marketing channels can be organised such that small-scale primary producers from developing economies are integrated into international marketing channels in a way that adds to the profitability of both the chain members and the welfare of the local communities, without compromising the sustainability of natural resources. She presented papers at several international conferences, for example: "Stretching Corporate Social Responsibility Upstream: Improving Sustainability with Upstream Partners in Global Marketing Channels" (Conference on Corporate Responsibility and Global Business: Implications for Corporate and Marketing Strategy, London Business School, United Kingdom, July, 13-14, 2006). She contributed a chapter titled *"Improving Quality and Ecological sustainability for Natural Resources in International Supply Chains: The Role of Market-Based Incentives,"* in the book *Integrated Agri-food Chains and Networks; Management and Organization* (Wageningen Academic Publishers, 333-342). Her research interests are in the area of marketing, channel governance, natural resources management notably in the context of small-scale producers in developing economies.

Contact: ekambewa@hotmail.com

Printed in the United States
by Baker & Taylor Publisher Services